Crystal Defects and Crystalline Interfaces

W. Bollmann

Crystal Defects and Crystalline Interfaces

With 158 Figures
and a Set of Moiré-Models

Springer-Verlag Berlin Heidelberg GmbH

W. BOLLMANN, Dr. sc. nat., Dipl. Phys., Battelle Institute, Advanced Studies Center, Geneva, and Privatdozent, Eidgenössische Technische Hochschule, Zürich, Switzerland

ISBN 978-3-642-49175-7 ISBN 978-3-642-49173-3 (eBook)
DOI 10.1007/978-3-642-49173-3

Illustration on the dust cover: Dislocations and grain boundaries in stainless steel, taken with the 1500 kV electron microscope at Toulouse, France, by G. Dupouy and F. Perrier.

Originally published by Springer-Verlag Berlin Heidelberg in 1970
Softcover reprint of the hardcover 1st edition 1970

 Printed by Universitätsdruckerei H. Stürtz AG, Würzburg. Title No. 1700.

To my Mother and my Wife

Preface

It is normal for the preface to explain the motivation behind the writing of the book. Since many good books dealing with the general theory of crystal defects already exist, a new book has to be especially justified, and here its main justification lies in its treatment of crystalline interfaces. About 1961, the work of the author, essentially based on the fundamental work of Professor F. C. Frank, started to branch away from the main flow of thought in this field and eventually led to a general geometrical theory which is presented as a whole for the first time in this book. Although nearly all that is presented has already been published in different journals and symposia, it might be difficult for the reader to follow that literature, as a new terminology and new methods of analysis had to be developed.

Special emphasis is given to discussion and many diagrams are included in order that a clear view of the basic concepts be obtained. Intermediate summaries try to bring out the main points of the chapters. Instead of specific exercises, general suggestions for them are given. The part up to chapter 9 is considered more or less as introductory, so that the book can be studied without specific knowledge of crystals and crystal defects. The presentation of that part developed out of lectures given by the author at the Swiss Federal Institute of Technology (ETH) in Zürich. As metallurgists and mineralogists are often not familiar with linear algebra, which is the mathematical basis of the interface theory, the procedures involved in this technique are collected in the Appendix. Many of the ideas contained in the book originated from observations made on moiré models and the subsequent formulation of these observations in mathematical terms. In order that the reader may participate in this experience, a set of such models is presented with the book.

Now the work is finished, it is my great pleasure to express my gratitude first to the Battelle Institute (BI) which by its financial support made this work possible, and in particular to Dr. B. D. Thomas, former president of the Battelle Memorial Institute, Dr. H. Thiemann, General Director of the Battelle Laboratories, Geneva and Dr. F. I. Milford, Director of Physical Sciences of BI, for their moral support of this work. Much inspiration came from discussions with my friend and former colleague Prof. D. G. Brandon, and I would like to express my special thanks to him. I should also like to thank Mr. P. Fontaine, my collaborator for the last fourteen years, who prepared many of the diagrams appearing in this book, including the precision drawings of the moiré models, Dr. and Mrs. E. Anderson for their contribution to the translation of the original German manuscript and Mrs. R. Anderson for the typing of this manuscript. Finally, I should like to thank the many others, not mentioned by name here, without whose contributions the book would not have become what it is, especially the publisher and his collaborators.

Geneva, July 1970 W. Bollmann

Contents

1. Introduction

1.1. The Significance of Crystal Defects

On opening a book about crystal defects the reader may wonder why this field should be investigated at all. It is well known that most of the solid materials exist in crystalline form, whether this fact is directly evident as in minerals or whether it becomes apparent only after preparation (polishing, etching) as in metals. Crystals are distinguished by their periodic, highly symmetric structures which possess a strong aesthetic appeal. The word "defect", however, indicates a disturbance, an imperfection, which, at first sight, appears to be neither aesthetic nor attractive.

The final aim of research is to attain an understanding of the objective, in this case of solid matter. Here it was seen quite early that the assumption of a perfect crystal led to results differing greatly from the measured data, especially with respect to flow stress, where the calculated results were higher than the measured data by a factor 100. Such observations led to the assumption that defects were already present in crystals. An understanding of the "real crystal" (in contrast to the "ideal crystal") is unthinkable without the knowledge of crystal defects. Crystal defects were investigated theoretically long before they could be made directly visible. We do not enter the history of the research on crystal defects as this is given in the books of Nabarro (1967) and Hirth and Lothe (1968).

1.2. The Nature of Crystal Defects

In order to become acquainted with the nature of crystal defects we first have to consider their environment, i.e. the crystal itself (Chapters 2–3).

By neglecting lattice oscillations, the crystal defects can be divided into the following groups, according to their dimensional nature:

a) Point defects: Vacancies, interstitial atoms, impurity atoms, and combinations of these;

b) line defects: i.e. dislocations;

c) two-dimensional defects: stacking faults, grain boundaries, etc.;

d) three-dimensional defects: precipitates, inclusions, holes, etc.

The central element of our study is the *line defect* or the dislocation. The theory of point defects is essentially independent of the theory of line defects, and the methods of investigation of these defects also differ.

After a brief discussion of point defects (Section 4.1) the book deals with the individual dislocation (5–6), the interactions between two dislocations (7–9), the interactions between dislocations and point defects (10) and the formation of dislocation networks (11). The second part of the book treats in detail the general geometrical theory of crystalline interfaces (12–14) which may be either boundaries between crystals of the same nature (grain boundaries) or boundaries between crystals of different types (phase boundaries). The applications show that this theory, which is based on purely geometrical concepts, already leads to significant physical information, and we believe that it will make a sound foundation for a wide field of physical research into intercrystalline boundaries.

The summaries try to provide a general picture of a group of chapters and, in this manner, to discuss the most important points. Occasionally it might be advantageous to read the summary before studying the detailed text.

As the subject matter for this book has originated from the direct electron microscope observation of crystal defects, emphasis is placed more on the static than on the dynamic aspects. The problems of dislocation dynamics, i.e. the theory of plastic deformation, is mentioned only briefly (10.2).

2. General Aspects of the Structure of Crystals

2.1. Outline

Solid state matter can be considered theoretically from different view-points depending on the range of problems to be investigated. For example, the elastic behaviour is generally discussed in terms of the continuum aspect. Other properties can be explained on the basis of the atomic arrangement. Our main interest concerns problems related to the plastic behaviour of materials and, for dealing with those, we need a certain knowledge of the atomic structure.

Concerning the atomic order two extreme states are distinguished, the amorphous state, which is mainly disordered (e.g. glass) and the essentially ordered crystalline state. In reality a certain local order prevails even in the amorphous state. The characteristic feature of a material in the crystalline state—i.e. of a "crystal"—is the periodicity of its structure. A crystal consists of identical elements (building blocks) which are joined together additively.

We may distinguish one-, two- and three-dimensional crystals. Examples of one-dimensional crystals are chain molecules such as nylon or silicate fibers. These molecules can, of course, assemble to form three-dimensional arrangements; however, the bonds along the individual fibers are much stronger than the inter-fiber bonds with the result that, in solution nylon separates into single molecular chains (Fig. 2.1/1).

An example of a two-dimensional crystal is graphite (see Section 3.8). Within the layers the bonds are covalent and therefore very strong, while the interlayer bonds are of the weak Van der Waals type. Other materials, which cleave easily, such as mica, can also be understood as being made up of stacked layers of two-dimensional crystals (Fig. 2.1/2). This concept of the two-dimensional crystal is even useful for the understanding of certain three-dimensional crystals (Section 3.2).

The word "crystal" usually means a three-dimensional crystal already distinct by its external shape (e.g. the tungstane crystals of Fig. 2.1/3). The external shape is—as is well known—a consequence of the internal structure and most substances are crystalline without this fact being evident from the exterior.

We have to distinguish clearly between the "model" of a crystal, its abstract image, i.e. the *ideal crystal*, and the real substance, i.e. the

Fig. 2.1/1. Disentangled nylon fibres as an example of a composite of one-dimensional crystals

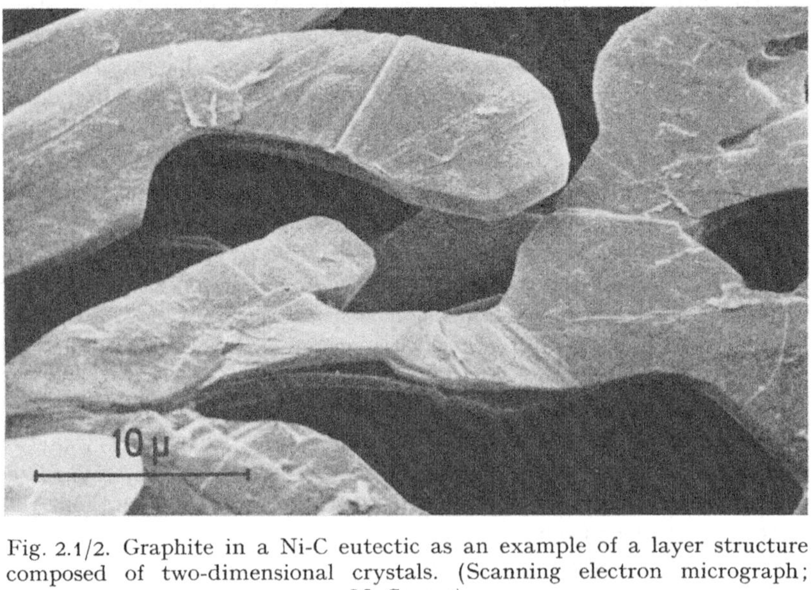

Fig. 2.1/2. Graphite in a Ni-C eutectic as an example of a layer structure composed of two-dimensional crystals. (Scanning electron micrograph; M. Grages)

real crystal, with all its imperfections and variations. The ideal crystal represents only certain restricted aspects of the real material.

Before going into a discussion of certain specific crystal structures we shall sketch some principles of crystallography. It might be thought that the understanding of crystal defects would require a comprehensive knowledge of the perfect crystal. However, in reality only a very limited acquaintance with it is necessary.

Fig. 2.1/3. Tungsten produced by the gas phase decomposition of WS_2, as an example of three-dimensional crystals. (Scanning electron micrograph; M. Grages)

2.2. Mathematical Description of the Crystal

As already mentioned in Section 2.1, the decisive feature of a (three-dimensional) crystal is the periodic arrangement of the atoms (in three dimensions). If one were to imagine part of the atomic pattern cut out, then this cut out part could be moved by parallel displacement (i.e. by translation) in such a manner that it would coincide at the new place with the pattern already existing there. Suitable selection of the cut out part could lead to a situation where the whole pattern could be built up *without gaps left* by translation of the chosen part. In general, the smallest possible part in the form of a parallelepiped with which this construction can be carried out is called *elementary cell* or *unit cell.* Sometimes a larger elementary cell is chosen in order to bring certain symmetries to light (Fig. 2.2/1).

Fig. 2.2/1. Choice of different elementary cells; pattern designed by M. C. Escher

The crystal framework obtained by joining unit cells is called a *translation lattice**. The elementary cell of the three-dimensional translation lattice is given by three linearly independant translation vectors $\boldsymbol{t}_1$, $\boldsymbol{t}_2$ and $\boldsymbol{t}_3$. Starting from a chosen origin every position $\boldsymbol{x}$ inside the

* In crystallography the translation lattice which is a mathematical concept is simply called "lattice". However, as we deal here with crystal defects we would not be allowed to use this word as soon as a defect were present. Therefore we use the word "lattice" in a wider sense to designate a more or less regular point arrangement and, if necessary, we give special emphasis to the translation lattice.

crystal is determined by:

$$\boldsymbol{x} = x^1\, \boldsymbol{t}_1 + x^2\, \boldsymbol{t}_2 + x^3\, \boldsymbol{t}_3 . \tag{2.2-1}$$

The meaning of the upper and lower indices is explained in the Appendix A1. Vectors and tensors are indicated by small- and capital *boldface* letters respectively, and the coordinates by normal letters.

The translation vectors $\boldsymbol{t}_i$ in their turn can be characterized by the so-called *lattice constants*, i.e. the lengths of the three translation vectors $a\,(=|\boldsymbol{t}_1|)$, $b\,(=|\boldsymbol{t}_2|)$ and $c\,(=|\boldsymbol{t}_3|)$ and the angles between them $\alpha\,(=\sphericalangle\boldsymbol{t}_2, \boldsymbol{t}_3)$, $\beta\,(=\sphericalangle\boldsymbol{t}_3, \boldsymbol{t}_1)$ and $\gamma\,(=\sphericalangle\boldsymbol{t}_1, \boldsymbol{t}_2)$.

For quantitative studies we have to place the crystal into a coordinate system. Of all the possible coordinate systems two types are of particular interest: The first type is the orthogonal system with three mutually perpendicular unit vectors of equal length (e.g. 1 Å). The second type is the crystal coordinate system which is determined by three linearly independent translation vectors (if possible the smallest translation vectors of the lattice).

The orthogonal coordinates are more convenient for plotting the results. With crystal coordinates a new coordinate system would have to be traced for each individual case. The crystal coordinate system corresponds better to the physical situation, as all the lattice points are given by integer coordinates. In practice it is often necessary to change from one coordinate system to the other.

In the orthogonal coordinate system the vector $\boldsymbol{x}$ shall be represented by:

$$\boldsymbol{x} = \xi^1\, \boldsymbol{u}_1 + \xi^2\, \boldsymbol{u}_2 + \xi^3\, \boldsymbol{u}_3 := \xi^i\, \boldsymbol{u}_i \; ^* \tag{2.2-2}$$

where, due to the orthogonality of the unit vectors:

$$(\boldsymbol{u}_i\, \boldsymbol{u}_k) = \delta_{ik} \begin{cases} = 1 & \text{if} \quad i = k \\ = 0 & \text{if} \quad i \neq k . \end{cases} \tag{2.2-3}$$

The transition from the orthogonal to the crystal coordinate system is formulated as follows: At first we express the basic vectors of the crystal system $\boldsymbol{t}_i$ by the orthogonal unit vectors $\boldsymbol{u}_i$ (without decomposing them into coordinates):

$$\boldsymbol{t}_k = \boldsymbol{u}_i\, s^i_k . \tag{2.2-4}$$

---

* We use the Greek letters ξ and ξ' in order to emphasize that the same vector is represented in different coordinate systems. The letters x and x' stand for coordinates of different vectors in one single coordinate system. According to Einstein's notation $\xi^i\, \boldsymbol{u}_i$ means $\sum_{i=1}^{3} \xi^i\, \boldsymbol{u}_i$. The symbol $:=$ means: equal by definition.

The orthogonal coordinates ξ^i are then expressed by the crystal coordinates ξ'^i by means of the same tensor or matrix $\mathbf{S}$.

$$\xi^i = s^i_k (\xi')^k. \tag{2.2-5}$$

The following example may serve as an illustration: We assume that $\boldsymbol{t}_1$ coincides with $\boldsymbol{u}_1$, i.e. with the x-axis, that $\boldsymbol{t}_2$ is the xy-face diagonal of the unit cube and $\boldsymbol{t}_3$ the cube diagonal (Fig. 2.2/2), i.e.

$$\begin{aligned} \boldsymbol{t}_1 &= \boldsymbol{u}_1 \cdot 1 + \boldsymbol{u}_2 \cdot 0 + \boldsymbol{u}_3 \cdot 0 \\ \boldsymbol{t}_2 &= \boldsymbol{u}_1 \cdot 1 + \boldsymbol{u}_2 \cdot 1 + \boldsymbol{u}_3 \cdot 0 \\ \boldsymbol{t}_3 &= \boldsymbol{u}_1 \cdot 1 + \boldsymbol{u}_2 \cdot 1 + \boldsymbol{u}_3 \cdot 1. \end{aligned} \tag{2.2-6}$$

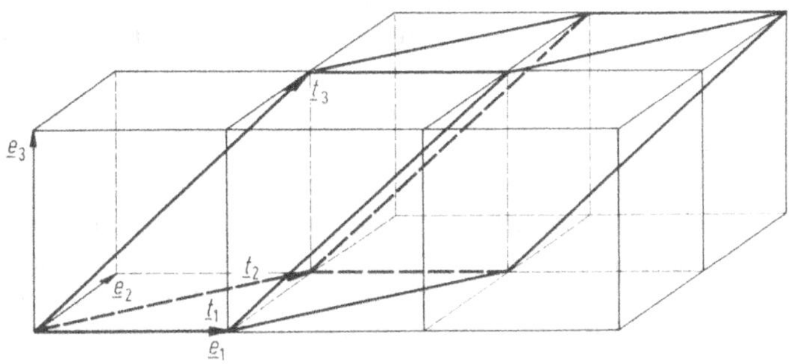

Fig. 2.2/2. Transition to a new elementary cell [Eq. (2.2-6)]

In matrix notation [Eq. (2.2-6)] is written:

$$(\boldsymbol{t}_1\, \boldsymbol{t}_2\, \boldsymbol{t}_3) = (\boldsymbol{u}_1\, \boldsymbol{u}_2\, \boldsymbol{u}_3) \begin{pmatrix} 1 & 1 & 1 \\ 0 & 1 & 1 \\ 0 & 0 & 1 \end{pmatrix} = \boldsymbol{u}_i\, s^i_k = \boldsymbol{t}_k. \tag{2.2-7}$$

The (vertical) column vectors of the $\mathbf{S}$-matrix are therefore the new basic vectors expressed in the coordinate system of the old ones. Thus, the transformation of the coordinates of an arbitrary vector is given by:

$$\xi^i = \begin{pmatrix} \xi^1 \\ \xi^2 \\ \xi^3 \end{pmatrix} = \begin{pmatrix} 1 & 1 & 1 \\ 0 & 1 & 1 \\ 0 & 0 & 1 \end{pmatrix} \begin{pmatrix} (\xi')^1 \\ (\xi')^2 \\ (\xi')^3 \end{pmatrix} = s^i_k (\xi')^k. \tag{2.2-8}$$

For example the vector $\boldsymbol{x}$ with the crystal coordinates $(\xi')^i = [0, 1, 0]$ is given in orthogonal coordinates $\xi^i = [1, 1, 0]$. Accordingly, the inverse matrix $\mathbf{S}^{-1}$ describes the transition from the orthogonal to the crystal coordinates.

$$S^{-1} = \begin{pmatrix} 1 & -1 & 0 \\ 0 & 1 & -1 \\ 0 & 0 & 1 \end{pmatrix} \tag{2.2-9}$$

$$(\xi')^i = (s^{-1})^i_k \, \xi^k \tag{2.2-10}$$

i.e.

$$\begin{pmatrix} (\xi')^1 \\ (\xi')^2 \\ (\xi')^3 \end{pmatrix} = \begin{pmatrix} 1 & -1 & 0 \\ 0 & 1 & -1 \\ 0 & 0 & 1 \end{pmatrix} \begin{pmatrix} \xi^1 \\ \xi^2 \\ \xi^3 \end{pmatrix}. \tag{2.2-11}$$

All the transitions from one coordinate system to another or from one vector system to another, described by matrices with constant terms as above are called "linear transformations". We distinguish between coordinate transformations and vector transformations. The formulation of the transformation matrix $\boldsymbol{S}$ for a general crystal with given lattice constants is indicated in the Appendix A1 Eq. (A1-26).

If we need to know the length of a vector or the angle between two vectors while those are expressed in an arbitrary coordinate system, this information can be acquired in two ways: The first consists in transforming the vectors into the orthogonal system where the required data are obtained by means of the usual scalar products:

$$|\boldsymbol{x}|^2 = (\boldsymbol{x}\,\boldsymbol{x}) \tag{2.2-12}$$

and

$$\cos(\sphericalangle \boldsymbol{x}, \boldsymbol{y}) = \frac{(\boldsymbol{x}\,\boldsymbol{y})}{|\boldsymbol{x}|\,|\boldsymbol{y}|} \tag{2.2-13}$$

(where $\boldsymbol{x}$ and $\boldsymbol{y}$ are expressed in orthogonal coordinates).

The more direct method, however, makes use of the metric tensor $\boldsymbol{G}$, which depends on the coordinate system

$$\boldsymbol{G} = \boldsymbol{S}^T \boldsymbol{S}. \tag{2.2-14}$$

($\boldsymbol{S}^T$ = transposed matrix of $\boldsymbol{S}$. See Appendix A1.)

In the general case with $\boldsymbol{x}'$ and $\boldsymbol{y}'$ in the arbitrary coordinates, this leads to

$$|\boldsymbol{x}'|^2 = (\boldsymbol{x}'^T \boldsymbol{G} \boldsymbol{x}') \tag{2.2-15}$$

and

$$\cos(\sphericalangle \boldsymbol{x}' \boldsymbol{y}') = \frac{(\boldsymbol{x}'^T \boldsymbol{G} \boldsymbol{y}')}{|\boldsymbol{x}'|\,|\boldsymbol{y}'|}. \tag{2.2-16}$$

The metric tensor is always symmetrical. For a cubic crystal coordinate system with the lattice constant a it becomes:

$$\boldsymbol{G} = \begin{pmatrix} a^2 & 0 & 0 \\ 0 & a^2 & 0 \\ 0 & 0 & a^2 \end{pmatrix}. \tag{2.2-17}$$

For the example (2.2-7), $\boldsymbol{G}$ is

$$\boldsymbol{G}=\begin{pmatrix}1 & 1 & 1\\ 1 & 2 & 2\\ 1 & 2 & 3\end{pmatrix}. \qquad (2.2\text{-}18)$$

Throughout this book we shall indicate vectors (translation vectors, Burgers vectors etc.) by their dimensionless coordinates. Their dimensions (length, angle, etc.) will be given by the metric tensor of the corresponding coordinate system.

It is necessary, on occasion, to choose new basis vectors in a crystal coordinate system such that the volume of the elementary cell remains unchanged. This is done by a so-called "*unimodular transformation*" $\boldsymbol{U}$. The fact that the volume stays unchanged is expressed by

$$\det(\boldsymbol{U})=\pm 1. \qquad (2.2\text{-}19)$$

The group of the linear transformations contains as a subgroup the unimodular transformations and those contain the symmetry operations, which leave the size and shape of the elementary cell unchanged (Section 2.3). The transformation $\boldsymbol{S}$ of the example (2.2-7) is a unimodular transformation, yet not a symmetry operation.

In the general theory of crystalline interfaces two adjacent crystals of the same structure may need to be described by different elementary cells. For this reason the relations between different possible elementary cells are emphasized here.

If we choose in a crystal coordinate system an elementary cell as unit cell*, then all the integer coordinates indicate discrete *points of the translation lattice*. We name these points "lattice points" A lattice point is the origin of the corresponding unit cell. Non-integer coordinates indicate points *within* a unit cell (which may or may not be atomic positions).

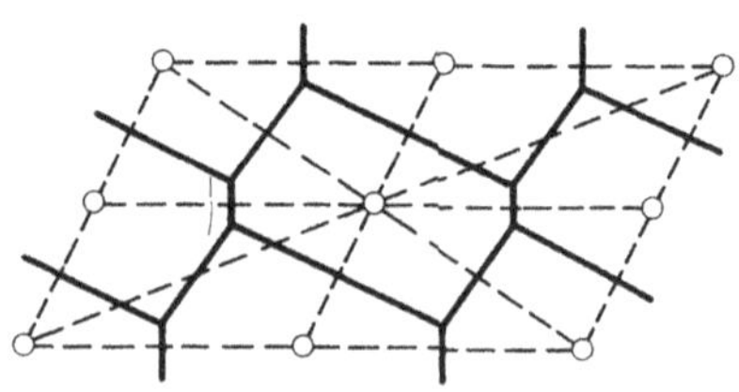

Fig. 2.2/3. Two dimensional example of Wigner-Seitz cells

* In this connection we understand as elementary cell the smallest part of the crystal pattern (of atomic positions) in the shape of a parallelepiped through which the complete pattern of the crystal can be obtained by translation. The unit cell, however, is described by the units of the coordinates.

Up to now we have assumed that the elementary cell is of the shape of a parallelepiped. However, other shapes are conceivable. Of particular interest is the '*Wigner-Seitz cell*' which is delimited by planes perpendicularly bisecting the straight line segments between two nearest neighbour lattice points. In this manner every lattice point is placed at the centre of a polyhedron (Fig. 2.2/3). We shall have to deal with Wigner-Seitz cells in connection with dislocation networks (Section 12.12).

2.3. Classification of Crystal Structures

We restrict ourselves in this Section to a few remarks. More detailed descriptions are to be found in all standard textbooks on crystallography.

All types of crystals are divided into seven *crystal systems* according to the shape of the elementary cell, whereas this latter is chosen such as to show the highest symmetry of the system.

Crystal system	Lattice constants
Triclinic	$a \neq b \neq c,\ \alpha \neq \beta \neq \gamma$
Monoclinic	$a \neq b \neq c,\ \alpha = \beta = 90^\circ,\ \gamma \neq 90^\circ$
Orthorhombic (Rhombic)	$a \neq b \neq c,\ \alpha = \beta = \gamma = 90^\circ$
Tetragonal	$a = b \neq c,\ \alpha = \beta = \gamma = 90^\circ$
Cubic	$a = b = c,\ \alpha = \beta = \gamma = 90^\circ$
Rhombohedral (Trigonal)	$a = b = c,\ \alpha = \beta = \gamma \neq 90^\circ$
Hexagonal	$a = b \neq c,\ \alpha = \beta = 90^\circ,\ \gamma = 120^\circ$

The cubic system is a limiting case of both the tetragonal and the rhombohedral systems. The rhombohedral system itself can be interpreted as a special case of the hexagonal system.

All possible translation lattices are divided into 14 Bravais lattices. The first group of seven is identical with the seven crystal systems while the second group is derived from the first by adding face centres, base centres and body centres.

The internal order is specified by symmetry operations. Those are transformations which recreate the same point arrangement as the initial one. Translations by translation vectors are specific symmetry operations. A special class are the *point symmetry operations* which leave at least one point (i.e. the origin) fixed and which are represented by homogenious linear transformations (see Appendix A 1). Point symmetry operations are rotations by 360° (1-fold rotation axis), 180° (2-fold), 120° (3-fold), 90° (4-fold) and 60° (6-fold), inversion at a point, reflection at plane and all the possible combinations of these operations. In this way 32 different *point groups* are distinguished.

230 space groups result from the combinations of the point symmetry operations with translations. Space symmetry elements are screw axes and glide reflection planes. Those operations are represented by non-homogenious linear transformations (Appendix A1).

In the next chapter we shall discuss in greater detail several special crystal structures.

3. Some Special Crystal Structures

3.1. Outline

In this chapter we shall discuss several common crystal structures, some of which are of prime importance in metallurgy—such as the face-centred cubic structure (austenitic steels, aluminium, nickel, copper, silver, gold, etc.), the close packed hexagonal structure (zinc, cobalt and others) and the body-centred cubic structure (ferritic steels, chromium, molybdenum, tungsten, etc.). The diamond structure prevails in the field of semiconductors, while (the hexagonal) graphite is important as a moderator material in reactor technology and as a precipitate in cast-iron.

Moreover, we shall consider different types of imperfections related to the various structures.

3.2. Close-Packed Structures

The basic element of the so-called close packed structures is a planar layer of equal spheres in the densest possible arrangement. Every sphere represents an atom (Fig. 3.2/1), and as such is surrounded in the plane by six nearest neighbours. This kind of a layer of spheres can be interpreted as a two-dimensional hexagonal crystal. Various stacking arrangements of these layers lead to different three-dimensional structures.

We imagine a close packed layer to be given and the centres of the spheres to be projected on the basal plane. This results in a triangular (planar) point lattice. The set of these points defines the position of this layer which we term as position A. If a second layer is placed on top of the first one, *two different stable positions are possible* and these we call B and C. The spheres of the second layer fit into the hollows formed between the spheres of the first layer. The fact that the number of hollows between the spheres is twice the number of spheres accounts for the two different possible position B and C of the second layer. As long as we assume that the atoms are held together mainly by isotropic central forces, the position A is excluded for the second layer since the spheres would then be in labile positions (on top of one another).

With only two layers given the positions B and C are physically fully equivalent. They are transferred into one another by a rotation of 60° around a vertical axis through an A-point. Therefore we may choose arbitrarily for the second layer the position B or C, in other words, we name the chosen position B or C.

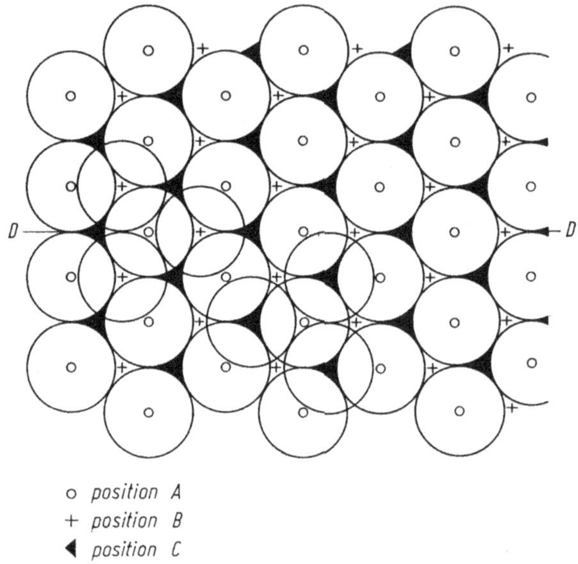

Fig. 3.2/1. A-, B- and C-positions in the close-packed arrangement of spheres

The position of the second layer shall now be called B. The third layer can again be placed in two positions, namely C and A. The physical situation differs, depending on the choice of that position. If we choose the position C, a sequence ABC is obtained. Relative to an atom in the B-position (in the second layer) all the direct neighbour atoms form a centrally symmetrical arrangement (Fig. 3.2/2a). However, if we chose position A for the third layer the sequence ABA is realized. In this case the nearest neighbours of a B-atom are arranged in mirror

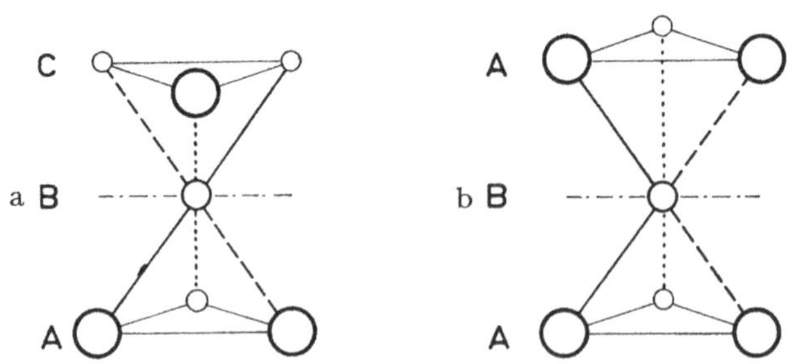

Fig. 3.2/2a and b. Nearest neighbours of an atom in the B-position, a) in the ABC-sequence, b) in the ABA-sequence

symmetry (Fig. 3.2/2b). For every subsequent layer a choice can be made between two of the three positions: *A*, *B* and *C*. These three positions are thus the only layer positions in close packed structures.

Of all the possible sequences of the positions *A*, *B* and *C* (whereby a sequence of the type *AA* is excluded) two are particularly important, namely ... *ABCABC* ..., the face-centred cubic structure and ... *ABABAB* ..., the close packed hexagonal structure. However, sequences of the type *AA* have been found in non close packed layer structures.

We introduce as an abbreviation (following Jagodzinski) the letter "*K*" ("cubic") for a sequence of the type *ABC* (*CBA*, *BAC*, etc.). In an analogous manner we denote by "*H*" ("hexagonal") a sequence of the type *ABA* (*BAB*, *CBC*, etc.). In this way we can describe the face-centred cubic structure by a series of *K*-sequences and the close packed hexagonal structure by *H*-sequences. For completeness we denote by "*L*" (labile) the (unstable) sequence of the type *AAB* (*ABB*, etc.).

The signs "▽" for the sequence *AB* (*BC*, *CA*) and "△" for *AC* (*CB*, *BA*) introduced by F. C. Frank, are often used in the literature. With this notation the *K*-sequence is expressed by ▽▽ or △△ and the *H*-sequence by △▽ or ▽△. We shall only use the *K*- and *H*-notation, which describes a situation that is invariant with respect to orthogonal transformations, while the △-sequence is inverted when the layer stack is rotated by 180° around a horizontal axis.

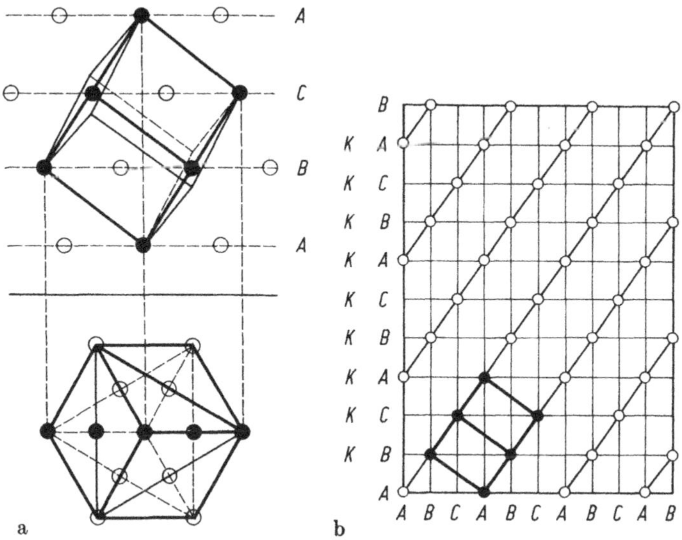

Fig. 3.2/3a and b. The face-centred cubic (f.c.c.) structure, a) projections, b) stacking sequence

From a vertical cut placed through the *D-D*-line in Fig. 3.2/1 result diagrams which are convenient for representing the stacking sequence such as Fig. 3.2/3 for the face-centred cubic and Fig. 3.2/4 for the close packed hexagonal structure. The *K*, *H* and *L*-sequences are indicated according to Fig. 3.2/5.

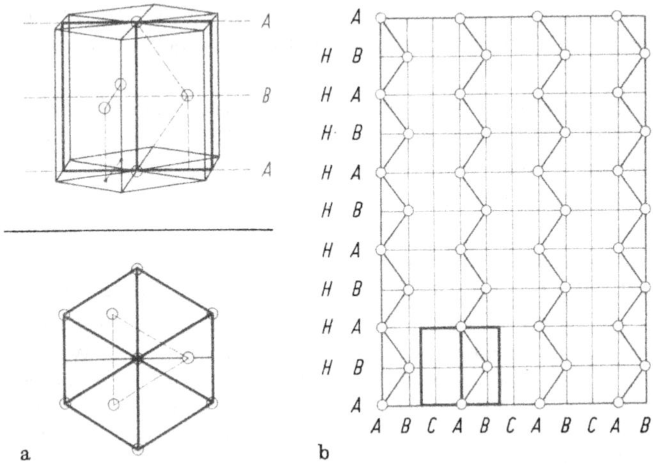

Fig. 3.2/4a and b. The close-packed hexagonal (c.p.h.) structure, a) projections b) stacking sequence

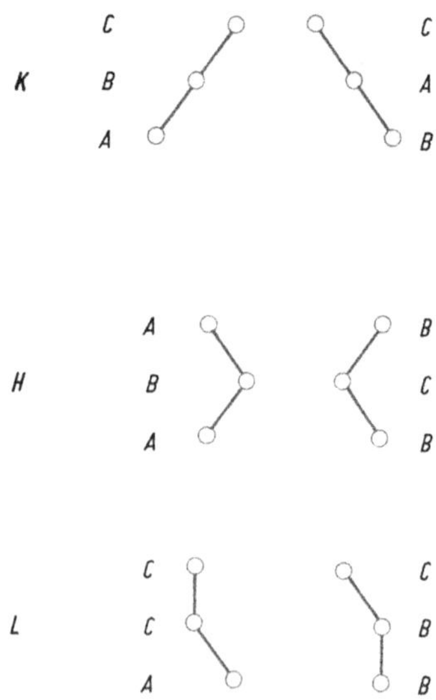

Fig. 3.2/5. *K*-, *H*- and *L*-sequences

3.3. The Face-Centred Cubic Structure

The elementary cell of the face-centred cubic structure (abbreviation: fcc-structure) is a cube with atoms at the corners and at the centres of the cube faces. It becomes evident that the fcc-structure corresponds to the ABC-stacking of close packed layers (Section 3.2) when the cube is set on one of its corners such that a cube-diagonal is vertical (Fig. 3.2/3).

Provided that we place the coordinate axes into the cube edges and choose the length of the cube edge as unit length, then the unit cell contains four atoms in the positions [000], [1/2, 1/2, 0], [1/2, 0, 1/2] and [0, 1/2, 1/2]* (Fig. 3.3/1). From the centrally symmetrical arrangement of the neighbour atoms around a given atom it follows that the same

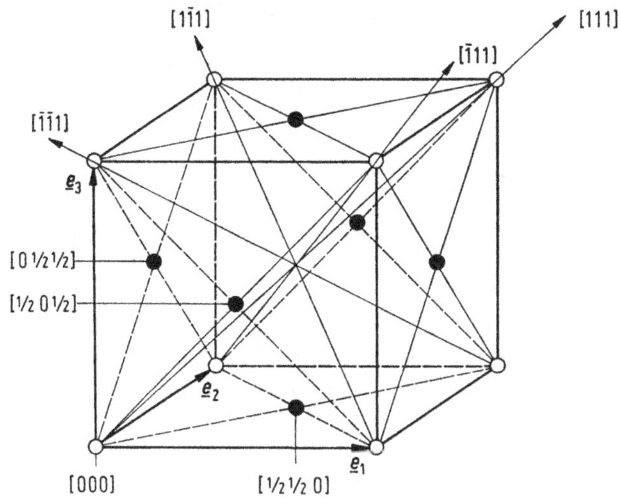

Fig. 3.3/1. Face-centred cubic elementary cell with atomic positions $e_i \equiv \boldsymbol{u}_i$

structure can be built by ABC-stacking in the four $\langle 111 \rangle$ directions ([111], [$\bar{1}$11], [1$\bar{1}$1], [$\bar{1}\bar{1}$1]**). The smallest translation vectors are of the type (1/2) $\langle 110 \rangle$, i.e. of half a surface diagonal of the elementary cube. The next larger translation vectors are of the $\langle 100 \rangle$-type. They may be composed of the smaller in the following way:

$$(1/2)\,[1\,1\,0] + (1/2)\,[1\,\bar{1}\,0] = [1\,0\,0]. \tag{3.3-1}$$

* Following a general convention we denote by the kind of brackets: special directions or vectors or positions by [...], the type of direction by $\langle \ldots \rangle$, special planes by (...), (where the coordinates are the reciprocal segments cut on the coordinate axes or the coordinates of the normal vector in the reciprocal lattice), and the type of the planes by {...}.

** According to the general convention $\bar{1} = -1$.

Vectors of the ⟨110⟩-type are even larger translation vectors; they result from:

$$(1/2)\ [1\,1\,0] + (1/2)\ [1\,1\,0] = [1\,1\,0] \tag{3.3-2}$$

or

$$[1\,0\,0] + [0\,1\,0] = [1\,1\,0]. \tag{3.3-3}$$

The fcc-structure can also be understood as a rhombohedral structure in which the (cubic) vectors [1/2, 0, 1/2], [1/2, 1/2, 0] and [0, 1/2, 1/2] are taken as unit vectors of the rhombohedral system (Fig. 3.3/2). Expressed

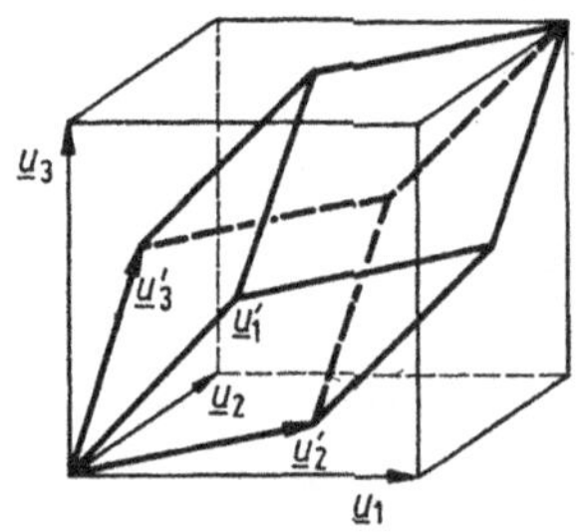

Fig. 3.3/2. Rhombohedral elementary cell of the f.c.c. structure

by the cubic unit vector $\boldsymbol{u}_i$ the rhombohedral unit vector $\boldsymbol{u}'_i$ becomes (Section 2.2):

$$\boldsymbol{u}_i\, s^i_k = \boldsymbol{u}'_k \tag{3.3-4}$$

i.e.

$$(\boldsymbol{u}_1\, \boldsymbol{u}_2\, \boldsymbol{u}_3) \begin{pmatrix} 1/2 & 1/2 & 0 \\ 0 & 1/2 & 1/2 \\ 1/2 & 0 & 1/2 \end{pmatrix} = (\boldsymbol{u}'_1\, \boldsymbol{u}'_2\, \boldsymbol{u}'_3). \tag{3.3-5}$$

The rhombohedral *coordinates* ξ'^i of a vector are then expressed in terms of the cubic ξ^i as follows:

$$\xi'^i = (s^{-1})^i_k\, \xi^k \tag{3.3-6}$$

i.e.

$$\begin{pmatrix} \xi'^1 \\ \xi'^2 \\ \xi'^3 \end{pmatrix} = \begin{pmatrix} 1 & 1 & 1 \\ 1 & 1 & -1 \\ -1 & 1 & 1 \end{pmatrix} \begin{pmatrix} \xi^1 \\ \xi^2 \\ \xi^3 \end{pmatrix}. \tag{3.3-7}$$

The volume of the rhombohedral unit cell is one quarter of the volume of the cubic unit cell $(\det(\boldsymbol{S}) = 1/4)$. If the value of the cubic lattice constant is a, the metric tensor of the rhombohedral coordinate system is given by:

$$\boldsymbol{G} = a^2 \begin{pmatrix} 1/2 & 1/4 & 1/4 \\ 1/4 & 1/2 & 1/4 \\ 1/4 & 1/4 & 1/2 \end{pmatrix}. \tag{3.3-8}$$

The rhombohedral unit cell contains only one atom, in contrast to the four atoms of the fcc-cell. The cubic symmetry is not immediately evident in the rhombohedral structure. However, four equivalent rhombohedral elementary cells can be defined in the direction of the four ⟨1 1 1⟩-axes. On the other hand, it becomes clear from the rhombohedral elementary cell, that the fcc-structure can be understood as a pure translation lattice.

3.4. The NaCl-Structure

As is well known, rocksalt is composed of Na^+- and Cl^--ions. If there were no difference between these ions the structure would be simple

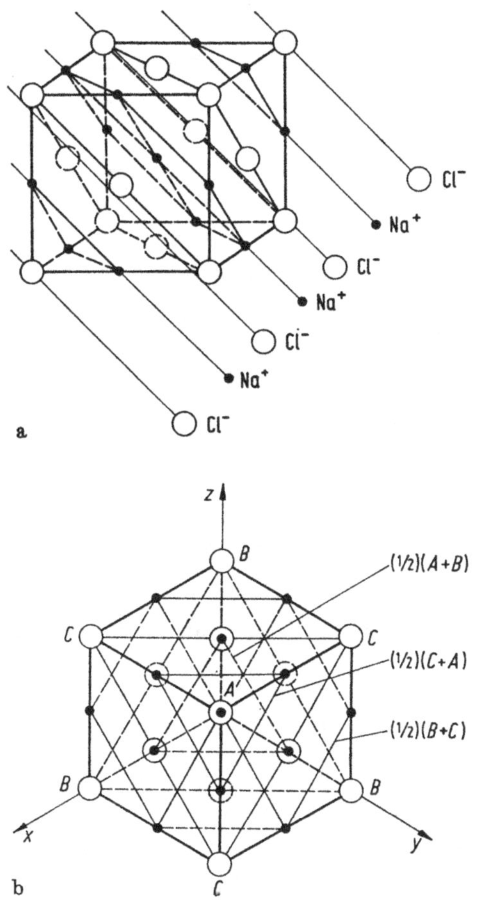

Fig. 3.4/1 a and b. Elementary cell of the NaCl structure a) parallel to individual ion layers, b) projection in the [1 1 1]-direction

cubic. However, as they differ, we pay attention, at the moment only, to one, e.g. the Cl^--ions. For these ions alone the structure is face-centred cubic. The Cl^--ions are located in the complete crystal in the standard positions: [000], [1/2, 1/2, 0], [1/2, 0, 1/2] and [0, 1/2, 1/2] (Section 3.3), while the positions of the Na^+-ions are: [1/2, 0, 0], [0, 1/2, 0], [0, 0, 1/2] and [1/2, 1/2, 1/2] (Fig. 3.4/1). The complete unit cell thus contains 8 atoms.

The Cl^--ions are arranged in a series of ABC-sequences (Section 3.2). Between every two Cl^--layers an Na^+-layer is located in the positions $(1/2)(A+B)$, $(1/2)(B+C)$ and $(1/2)(C+A)$, where A, B and C represent positions of nearest-neighbour atoms in the adjacent layers. The NaCl-structure can thus be understood as two interpenetrating fcc-structures, translated by [1/2, 0, 0] with respect to one another.

The NaCl-structure can also be interpreted as a rhombohedral structure with a unit cell containing a Cl^--ion at [000] and a Na^+-ion at [1/2, 1/2, 1/2], i.e. at the centre of the unit cell (Fig. 3.4/2).

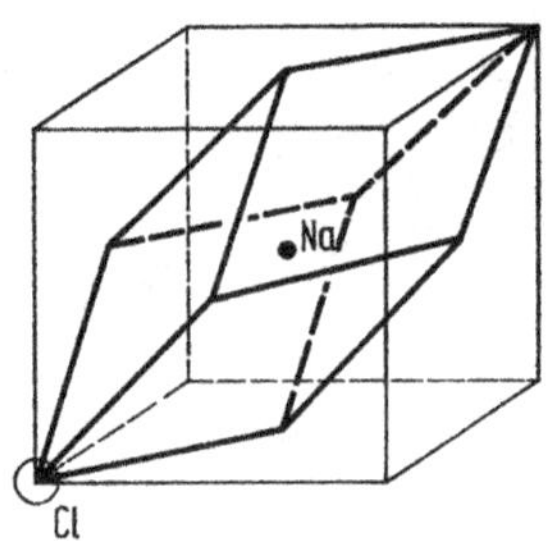

Fig. 3.4/2. Rhombohedral elementary cell of the NaCl structure

3.5. The Diamond Structure

The diamond structure may be understood as composed of a basic face-centred cubic framework with atoms in the standard positions [000], [1/2, 1/2, 0], [1/2, 0, 1/2] and [0, 1/2, 1/2] and a further group of atoms in the positions [1/4, 1/4, 1/4], [3/4, 1/4, 1/4], [3/4, 1/4, 3/4] and [3/4, 3/4, 3/4] (Fig. 3.5/1). The atoms in the two groups may be chemically identical as in diamond (carbon), germanium and silicon or chemically different as in the III-V-compositions such as ZnS, GaAs, etc. As in the NaCl structure the diamond unit cell contains also 8 atoms arranged in two interpenetrating fcc-lattices except that here the translation between the two lattices is [1/4, 1/4, 1/4].

The diamond structure can also be described as rhombohedral with a unit cell containing two atoms in the positions [000] and [1/4, 1/4, 1/4]

([3/4, 1/4, 1/4] respectively, depending on the choice of the unit vectors). The translation vectors of the diamond structure are therefore those of the fcc-lattice.

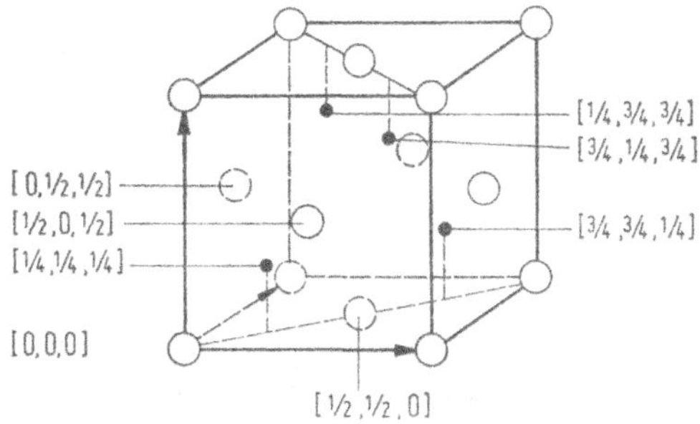

Fig. 3.5/1. Elementary cell of the diamond structure

Instead of a close packed structure of spheres, the diamond structure can be interpreted as a loose stacking of tetrahedra with atoms at the corners and at the centre of the tetrahedron (Fig. 3.5/2). Considering layers of single atoms, their stacking sequence can be written: ... $AA'BB'CC'AA'$ .... Here $AA'BB'CC'$ becomes a K-sequence and $AA'BB'AA'$ an H-sequence.

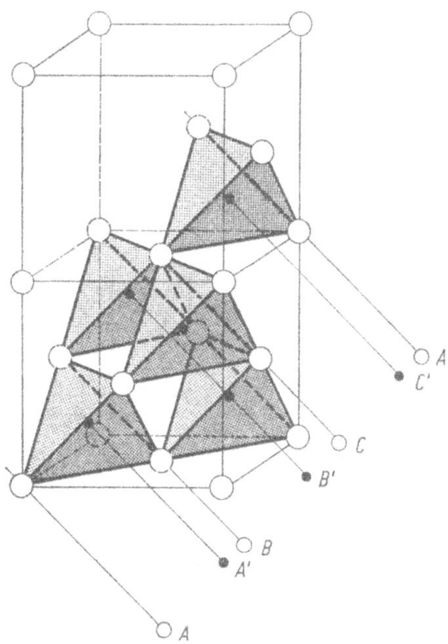

Fig. 3.5/2. The diamond structure represented as the stacking of tetrahedra

3.6. The Body-Centred Cubic Structure

The body-centred cubic structure (abbreviation: bcc-structure) is certainly one of the most important in metallurgy, since iron is crystallized in this structure at room temperature, as are other metals such as the alkali metals, V, Nb, Ta, Cr, Mo, W, etc.

The body-centred cubic unit cell contains two atoms in the positions [000] and [1/2, 1/2, 1/2] (Fig. 3.6/1). This structure is a translation

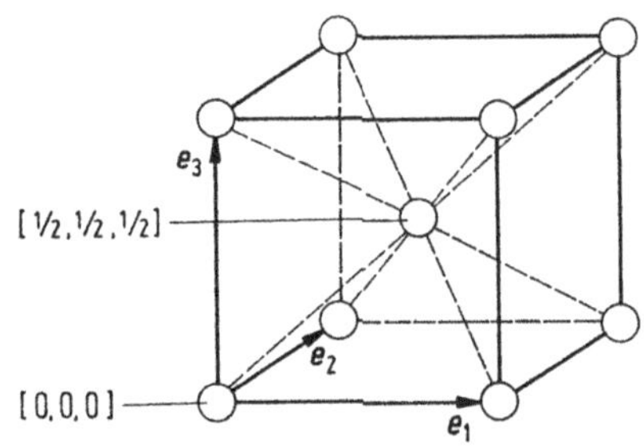

Fig. 3.6/1. The body-centred cubic elementary cell

lattice (Section 2.2) for which the following translation vectors, expressed in cubic coordinates, can be chosen (Fig. 3.6/2):

$$\begin{aligned} \boldsymbol{t}_1 &= [1/2 \quad 1/2 \quad 1/2] \\ \boldsymbol{t}_2 &= [-1/2 \quad 1/2 \quad 1/2] \\ \boldsymbol{t}_3 &= [0 \quad 0 \quad 1]. \end{aligned} \tag{3.6-1}$$

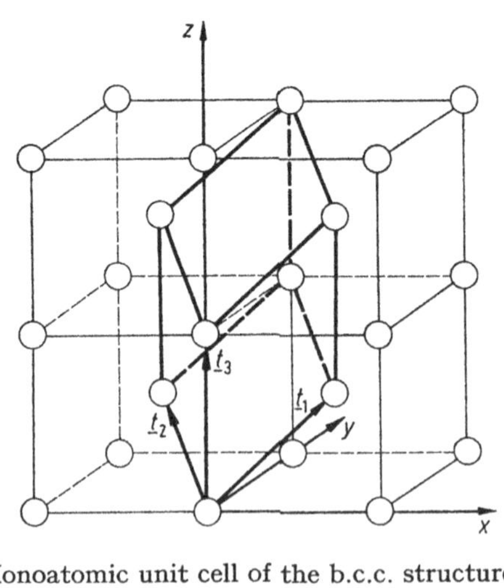

Fig. 3.6/2. Monoatomic unit cell of the b.c.c. structure [Eq. (3.6-1)]

From this follows the matrix $\boldsymbol{S}$ [compare Eq. (2.2-4)]:

$$\boldsymbol{S}=\begin{pmatrix}1/2 & -1/2 & 0\\ 1/2 & 1/2 & 0\\ 1/2 & 1/2 & 1\end{pmatrix} \tag{3.6-2}$$

which transforms the cubic into the oblique angled *unit vectors*. The matrix $\boldsymbol{S}^{-1}$

$$\boldsymbol{S}^{-1}=\begin{pmatrix}1 & 1 & 0\\ -1 & 1 & 0\\ 0 & -1 & 1\end{pmatrix} \tag{3.6-3}$$

transforms the cubic *coordinates* of a given vector into its oblique angled ones (2.2-10). The metric tensor of this new coordinate system is:

$$\boldsymbol{G}=a^2\begin{pmatrix}3/4 & 1/4 & 1/2\\ 1/4 & 3/4 & 1/2\\ 1/2 & 1/2 & 1\end{pmatrix}. \tag{3.6-4}$$

The new unit cell contains one atom and therefore its volume is half that of the cubic unit cell $(\det(\boldsymbol{S})=1/2)$. Its faces are $\{1\,1\,0\}$ planes. The shortest translation vectors of the bcc-structure are of the $(1/2)\langle 1\,1\,1\rangle$ type and the next longer ones of the $\langle 1\,0\,0\rangle$-type,

$$(1/2)\,[1\,1\,1]+(1/2)\,[\bar{1}\,\bar{1}\,1]=[0\,0\,1] \tag{3.6-5}$$

and the next of the $\langle 1\,1\,0\rangle$-type

$$(1/2)\,[1\,1\,1]+(1/2)\,[1\,1\,\bar{1}]=[1\,1\,0]. \tag{3.6-6}$$

Another primitive elementary cell can be defined by the following translation vectors (Fig. 3.6/3):

$$\begin{aligned} \boldsymbol{t}_1' &= [1/2 \;\; -1/2 \;\; 1/2]\\ \boldsymbol{t}_2' &= [1/2 \;\; 1/2 \;\; 1/2]\\ \boldsymbol{t}_3' &= [-1/2 \;\; 1/2 \;\; 1/2] \end{aligned} \tag{3.6-7}$$

with $\boldsymbol{S}$:

$$\boldsymbol{S}=\begin{pmatrix}1/2 & 1/2 & -1/2\\ -1/2 & 1/2 & 1/2\\ 1/2 & 1/2 & 1/2\end{pmatrix} \tag{3.6-8}$$

and $\boldsymbol{S}^{-1}$:

$$\boldsymbol{S}^{-1}=\begin{pmatrix}0 & -1 & 1\\ 1 & 1 & 0\\ -1 & 0 & 1\end{pmatrix}. \tag{3.6-9}$$

Here two faces of the unit cell are {1 1 1}-planes and one is a {1 1 0}-plane. The metric tensor is:

$$G = a^2 \begin{pmatrix} 3/4 & 1/4 & -1/4 \\ 1/4 & 3/4 & 1/4 \\ -1/4 & 1/4 & 3/4 \end{pmatrix}. \tag{3.6-10}$$

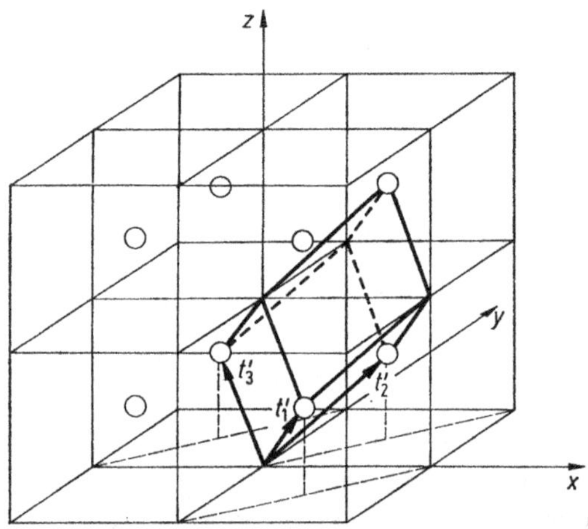

Fig. 3.6/3. Monoatomic unit cell of the b.c.c. structure [Eq. (3.6-7)]

The bcc-lattice can also be described by a face-centred tetragonal elementary cell (Fig. 3.6/4) by the new unit vectors:

$$\begin{aligned} \boldsymbol{t}_1'' &= [1 \quad 1 \quad 0] \\ \boldsymbol{t}_2'' &= [-1 \quad 1 \quad 0] \\ \boldsymbol{t}_3'' &= [0 \quad 0 \quad 1]. \end{aligned} \tag{3.6-11}$$

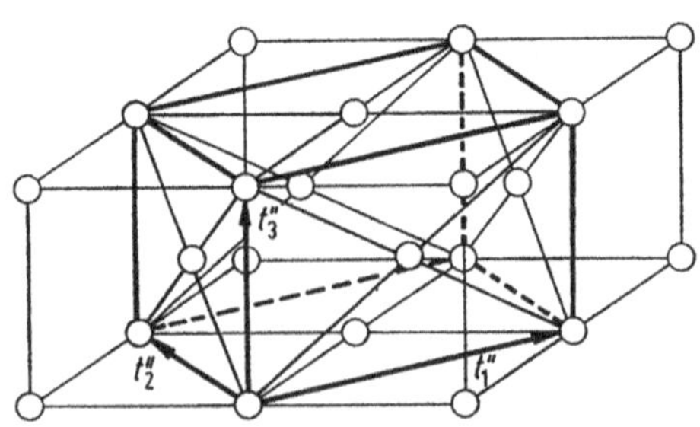

Fig. 3.6/4. Face-centred tetragonal elementary cell of the b.c.c. structure [Eq. (3.6-11)]

With the matrix $\boldsymbol{S}$:

$$\boldsymbol{S} = \begin{pmatrix} 1 & -1 & 0 \\ 1 & 1 & 0 \\ 0 & 0 & 1 \end{pmatrix} \tag{3.6-12}$$

and $\boldsymbol{S}^{-1}$:

$$\boldsymbol{S}^{-1} = \begin{pmatrix} 1/2 & 1/2 & 0 \\ -1/2 & 1/2 & 0 \\ 0 & 0 & 1 \end{pmatrix} \tag{3.6-13}$$

and $\boldsymbol{G}$:

$$\boldsymbol{G} = a^2 \begin{pmatrix} 2 & 0 & 0 \\ 0 & 2 & 0 \\ 0 & 0 & 1 \end{pmatrix}. \tag{3.6-14}$$

The tetragonal unit cell is obviously twice as large as the bcc cell, since it contains four atoms.

The above description demonstrates the relationship between the body-centred and the face-centred cubic lattices. One lattice can, in principle, originate from the other by pure deformation. This type of phase change is called the Bain transformation after the American metallurgist Edgar C. Bain.

3.7. The Close-Packed Hexagonal Structure

The close-packed hexagonal structure (cph-structure) is formed by an $\ldots ABAB \ldots$-stacking sequence (i.e. by a series of H-sequences) of close-packed layers of spheres (as already indicated in Section 3.2). Here the atoms are arranged mirror-symmetrically with respect to any close-packed layer-plane. The elementary cell can be defined by the three basic vectors $\boldsymbol{a}$, $\boldsymbol{b}$ and $\boldsymbol{c}$ (Fig. 3.7/1), where $\boldsymbol{a}$ and $\boldsymbol{b}$ lie in the basal plane, are of equal length (smallest distance between spheres) and enclose an angle of 120°, and $\boldsymbol{c}$ is perpendicular to $\boldsymbol{a}$ and $\boldsymbol{b}$ with a length equal to twice the interlayer spacing. The elementary cell thus defined contains two atoms, at the positions [000] and [1/3, 2/3, 1/2]. In the ideal cph-structure the ratio c/a is given by: $c/a = (2/3)\sqrt{6} = 1.630$. This becomes obvious when the positions of atoms are given in cubic coordinates, i.e. $|(2/3)\ [111]|/|(1/2)\ [110]|$. In real crystals the values can deviate considerably from the idealratio.

The close-packed hexagonal structure is not a translation lattice, i.e. not all the atomic positions can be obtained by composing basic vectors. If we assume the vector [1/3, 2/3, 1/2] pointing from the origin to the atom inside the elementary cell to be a translation vector, then another atom should be found at the position [−1/3, −2/3, −1/2], which is not the case. Therefore the translation lattice of the cph-structure is determined by $\boldsymbol{a}$, $\boldsymbol{b}$ and $\boldsymbol{c}$.

The metric tensor of the hexagonal structure is

$$G = \begin{pmatrix} a^2 & -a^2/2 & 0 \\ -a^2/2 & a^2 & 0 \\ 0 & 0 & c^2 \end{pmatrix}. \tag{3.7-1}$$

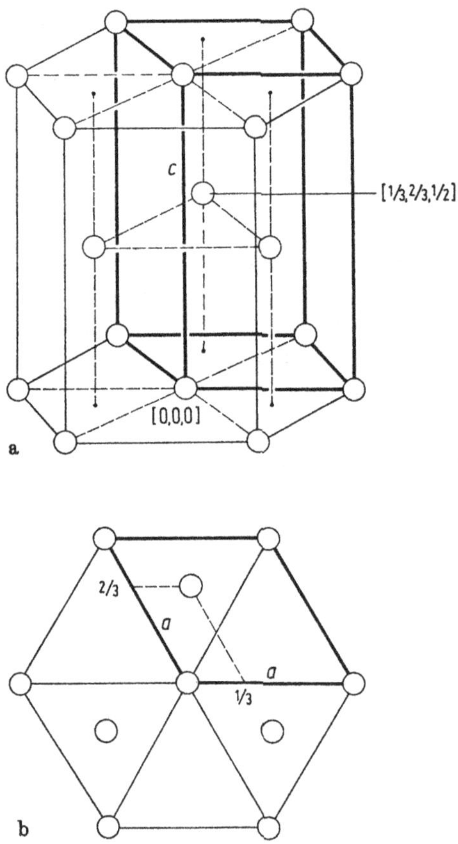

Fig. 3.7/1 a and b. The close-packed hexagonal structure; close-packed layers from a) side, and b) top

In the ideal cph-structure, c^2 is given by:

$$c^2 = (8/3)\, a^2. \tag{3.7-2}$$

In order to describe the hexagonal structure, a 4-coordinate-representations is often applied, in which the four coordinates refer to four unit vectors, three of which, lying in the basal plane, are rotated by 120° with respect to each other around the c-axis, and therefore are linearly dependant. This involves the additional condition that the sum of the first three coordinates must be zero. The unit vector [100] in this representation becomes:

$$t_1 = (1/3)\ [2\bar{1}\bar{1}0]. \qquad (3.7\text{-}3)$$

All the possible vectors of the same kind are obtained by permutation and change of sign of the first three coordinates; hence, the type of the vector is immediately evident.

We shall not use this representation, because it is removed from the general method. We shall use the methods of linear algebra which can be applied to any crystal system, and are convenient, in particular, in computer calculations. Furthermore, it might be necessary to choose for one given structure elementary cells of other systems (e.g. a rhombohedral instead of a hexagonal one), in which case a notation, exclusively valid for the hexagonal system, proves to be more of an obstacle than an advantage.

3.8. The Graphite Structure

The graphite structure, i.e. the structure of the carbon phase stable under normal conditions, is a very pronounced layer structure. The individual layers are formed from condensed benzene rings with the bonds between the atoms being homeopolar (Fig. 3.8/1). Between the layers the bonds are of the relatively weak Van der Waals type which accounts for the easy cleavage of these crystals.

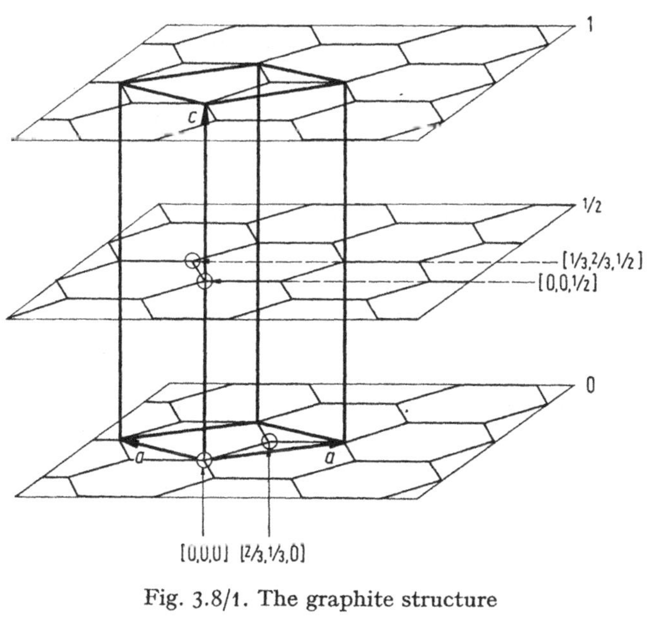

Fig. 3.8/1. The graphite structure

The natural graphite crystals are hexagonal in the—$ABAB$—stacking. The elementary cell contains four atoms in the positions: [000], [2/3, 1/3, 0], [0, 0, 1/2] and [1/3, 2/3, 1/2]. The metric tensor of the graphite structure is the same as that given in (3.7-1). The lattice constants are $a=b=2.46$ Å, $c=6.69$ Å.

3.9. Irregularities in the Stacking Sequence

In Sections 3.3 to 3.8, a number of structures which can be built up from the uniform stacking of atomic layers were discussed. In this section we shall investigate several kinds of "stacking faults" (i.e. planar faults) which originate from irregularities in the stacking sequence.

First we shall consider the face-centred cubic structure resulting from a series of K-sequences. If the crystal contains one single H-sequence in the interior, this means—as shown in Fig. 3.9/1—that it is twinned with

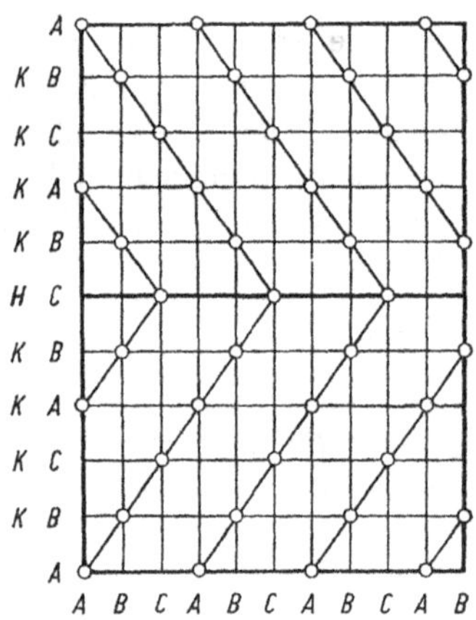

Fig. 3.9/1. Growth stacking fault (twin boundary)

the middle layer of the H-sequence being the *twin boundary*. This kind of stacking fault is generally called a *growth stacking fault*, because it occurs principally during crystal growth or recrystallisation. A series of two H-sequences is a so-called *deformation stacking fault* (Fig. 3.9/2). On speaking of stacking faults, deformation stacking faults are generally meant. The configuration of this type of fault can be produced by plastic deformation, i.e. by shifting one part of the crystal with respect to the other.

A translation vector for shifting a layer into an equivalent neighbouring position is of the $(1/2)\langle 110\rangle$-type, and for shifting it into a stacking fault position it is of the $(1/6)\langle 112\rangle$-type. Thus, a lattice translation in the (111)-plane can be split into two steps, e.g.

$$(1/2)\ [1\bar{1}0] = (1/6)\ [2\bar{1}\bar{1}] + (1/6)\ [1\bar{2}1]$$

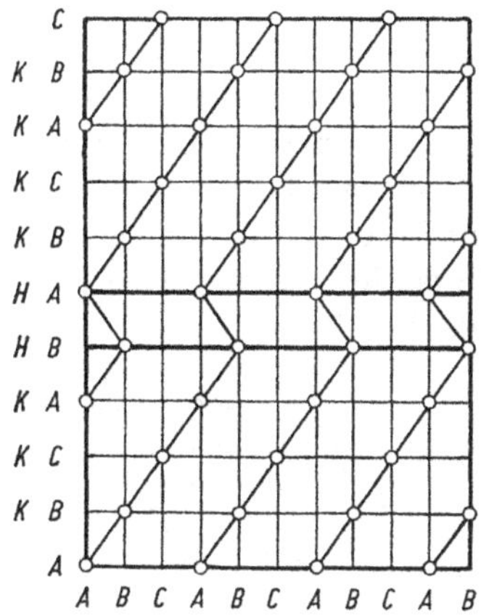

Fig. 3.9/2. Deformation stacking fault

where the first step leads from the perfect lattice position into the stacking fault position and the second step from this stacking fault position into the new perfect lattice position. The sequence of these steps cannot be chosen arbitrarily (see Section 8.2).

An important source of stacking faults is the condensation of vacancies or interstitial atoms. The condensation of vacancies corresponds to removing one layer followed by closure of the gap. We consider the sequence

$$\begin{gathered} \uparrow \\ \ldots ABCA(B)CABC \ldots \\ \ldots KKK\ K\ KKK \ldots. \end{gathered}$$

If the B-layer is removed we obtain

$$\begin{gathered} \ldots ABCA/CABC \ldots \\ \ldots KK\boldsymbol{H}\ \boldsymbol{H}KK \ldots \end{gathered}$$

i.e. the configuration of the deformation stacking fault (Fig. 3.9/2). Is, however, a layer added at the indicated location

$$\begin{gathered} \downarrow \\ \ldots ABCABCABC \ldots \\ \ldots KKKKKKK \ldots \end{gathered}$$

then it can occupy neither the A- nor the B-position because in these cases the unstable AA- or BB-sequence would result. Thus, the additional layer can only be placed in the *C*-position (Fig. 3.9/3).

... *ABCACBCABC* ...

... *KK***H***K***H***KKK* ...

According to F. C. Frank two kinds of stacking faults, called "intrinsic" and "extrinsic" faults are distinguished. At an intrinsic fault the crystal is undisturbed up to the boundary surface on approaching the disturbed area from either side [e.g. twin boundary or condensed vacancies (Figs. 3.9/1 and 2)]. An extrinsic fault, however, contains an intermediate layer which differs from the crystal lattice on both sides of the fault (condensed interstitials, Fig. 3.9/3).

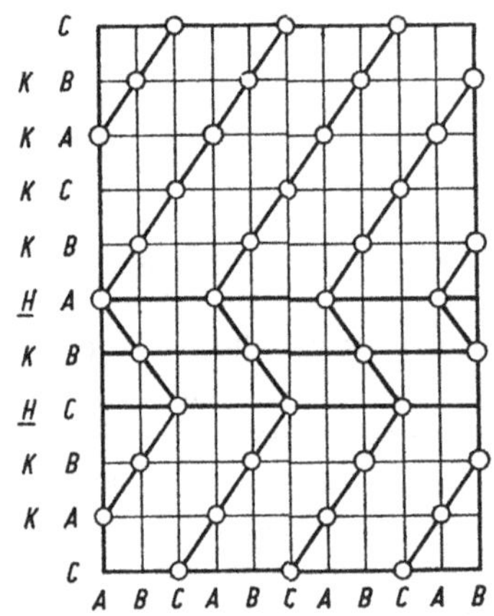

Fig. 3.9/3. Stacking sequence of condensed interstitial atoms

Stacking faults occur also in the hexagonal sequence. Condensed vacancies, e.g.

↑

... *ABAB(A)BABAB* ...

... *HHH H HHH* ...

produce, (Fig. 3.9/4a):

... *ABAB/BABAB* ...

... *HH***L** **L***HHH* ...

As mentioned above, this sequence is generally unstable, in particular if the atoms of the crystal are bonded by central forces. It can be stabilized by shifting one part of the crystal to one side or the other. For example:

... *ABAB***CBCBCB** ...

... *HH***K***HHHHH* ...

Here the stacking fault is intrinsic (Fig. 3.9/4b). For condensed interstitials the situation is:

↓

... *ABABABAB* ...

... *HHHHHH* ...

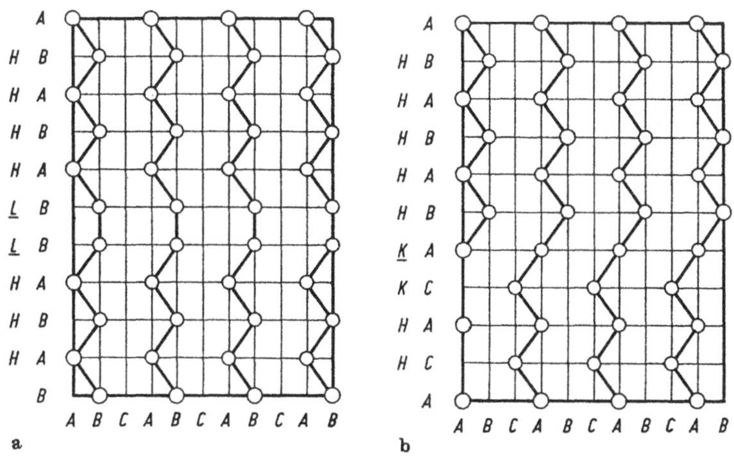

Fig. 3.9/4a and b. Stacking sequence of condensed vacancies in the c.p.h. structure, a) unstable sequence, b) stabilized sequence

Here a *C*-layer is, inserted (Fig. 3.9/5a. Is the drawing right?):

... *ABABCABAB* ...

... *HH**KKK**HH* ...

Thus, three consecutive *K*-sequences appear, which considerably disturb the hexagonal order. This arrangement can also be changed to a less disturbed by shifting one part of the crystal (Fig. 3.9/5b):

... *ABABCABABAB* ...

... *ABAB**ABCBCBC*** ...

... *HHHH**K**HHHH* ...

Now, only one *K*-sequence remains.

Considerations of this kind can be applied to the stacking sequences of other "layer structures" as well.

Also in the NaCl-structure certain kinds of stacking faults are possible on the {111}-planes, due to the opposite polarity (and hence attraction) of alternate layers. However, a shift by a (1/2) ⟨100⟩-type

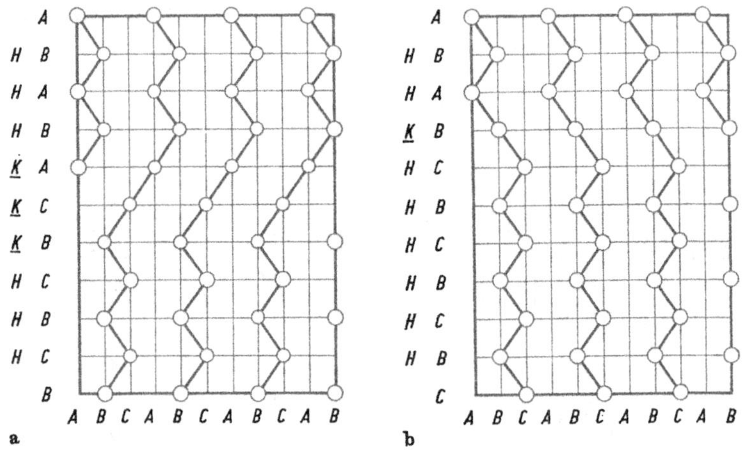

Fig. 3.9/5a and b. Stacking sequence of condensed interstitial atoms in the c.p.h. structure, a) high energy sequence, b) sequence with lower energy

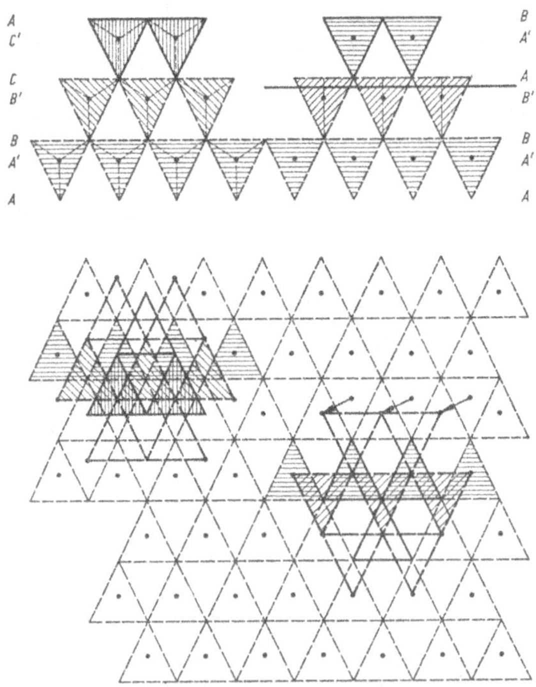

Fig. 3.9/6. Stacking faults in the diamond structure. (Identical positions in the front- and top views are marked by the same striated pattern)

vector in a $\{100\}$-plane brings atoms of the same polarity (repelling each other) together; this accounts for the $\{100\}$-cleavage facets of rock salt.

As the diamond structure can be considered as being built by the stacking of layers of tetrahedra, stacking faults analogous to those observed in the fcc-structure can also occur here. An H-sequence in the diamond structure can also be understood as being produced by a 60° rotation of the middle layer of a K-sequence around a vertical axis (Fig. 3.9/6).

In the graphite structure all the possible stacking faults of the hexagonal structure can occur, due to the weak interlayer bonds. Even the unstable AA-stacking can be found, e.g. in small areas of condensed vacancies (diameter up to several hundred Å). In this case the strong bonds within the layers prevent a lateral shift of the type illustrated in Fig. 3.9/4a.

A special kind of faulted layer structure is the *turbostratic* arrangement. In this case layers or layer packets, rotated by arbitrary angles with respect to one another, are stacked. Artificial graphite, used, for instance, as electrodes or as nuclear reactor moderators, is almost always found in this arrangement.

3.10. Energy of Stacking Faults

The fact that both face-centred cubic (e.g. Al, Au, Ag, Ni, austenitic steel, etc.) and close-packed hexagonal crystals (e.g. Co, Be, Mg, Cd) are found in nature shows that there must be a difference between the energies of the H- and the K-sequences. Otherwise both sequences would occur in a statistical distribution.

We now envisage an A-layer with a B-layer on top of it. For the third layer two possibilities exist: the A- or the C-position. Now we add the third layer. If we assume that the material crystallizes in the fcc-structure, this third layer must be attracted more strongly into the C- than into the A-position. The free energy of the crystal in the $ABC = K$-sequence is here necessarily lower than in the $ABA = H$-sequence. This energy has to be increased on shifting the third layer from the C- into the A-position (Fig. 3.10/1). The difference in free energy is obviously connected with the fact that, at the transition from a K- to an H-sequence, the arrangement of atoms changes from central symmetry to mirror symmetry. Although the bonding of close-packed structures is mainly through central forces, a small amount of bonding is directional. The energy difference ΔE between the K- and the H-sequence is a measure of the so-called "stacking fault energy" γ, with its value normally given in erg/cm^2. More exactly, γ in the fcc-structure indicates the energy per unit area of a deformation stacking fault, i.e. of a series of two H-sequences.

To a first approximation, the energies of different kinds of stacking faults can be estimated as being proportional to the number of foreign sequences (e.g. H-sequences in the fcc-structure). Thus, the energy of a twin boundary is approximately half that of a deformation stacking

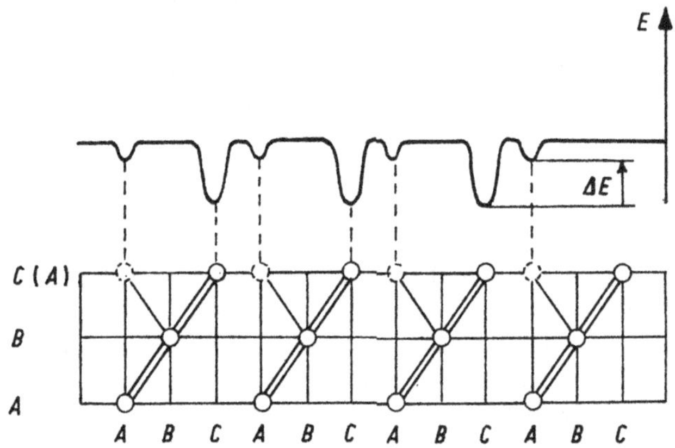

Fig. 3.10/1. Difference in the potential energy between the C- and the A-position (schematic representation)

fault. The energy of a layer of interstitial atoms in the hexagonal structure, as shown in Fig. 3.9/5, can thus be considerably lowered by shifting the layer.

These considerations show the advantage of the K- and H-notation. The values of stacking fault energies in the fcc-structure vary between practically zero (stainless steel) and ≈ 200 erg/cm^2 (Al). Numerical data are given in Hirth and Lothe (1968), p. 764 and Weertman (1964), p. 98.

3.11. Fibrous Structures

The case of graphite (Section 3.8) shows that the chemical bond can be different in different directions, thus producing strongly anisotropic materials. In layer structures the bond is stronger within the layers than between them. It is also possible that the bond in *one* direction is stronger than in all the others, thus creating primarily chain molecules with more or less loose side-bonds. Such arrangements are found in silicates and polymers (Section 2.1).

Several equivalent side bonds per unit length may be present, such that the chains can be linked to one another in different (relative) positions. An example for this is Tobermorite (Fig. 3.11/1), which plays an important part in the setting of cement. Other possibilities arise from the folding of chain molecules (polyethylene).

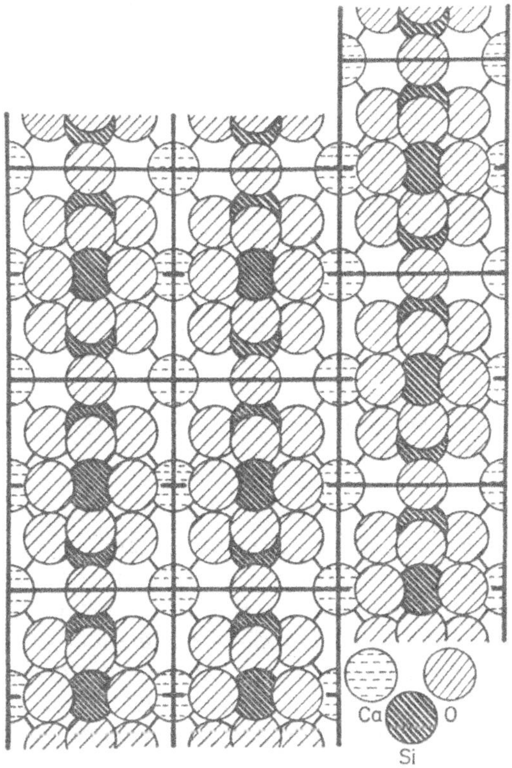

Fig. 3.11/1. Tobermorite as an example of a fibrous structure. (This drawing is based on the front cover of Scientific American, April 1964)

Chapters 2–3. Summary and Discussion

In describing the crystal we discussed principally its periodicity, i.e. the configurations produced by pure translations. Irregularities in the periodicity were also analysed. The point groups and space groups representing the symmetries of crystals, which are of prime importance in crystallography, were only briefly considered. These latter groups play a secondary part in the understanding of crystal defects, and, in particular, of crystalline interfaces; here elementary cells have often to be chosen, which do not show the highest symmetry of the crystal structure.

Some important structures were discussed, for example:

3.3. the face-centred cubic (fcc) structure,

3.4. the rock salt (NaCl) structure,

3.5. the diamond structure,

3.6. the body-centred cubic (bcc) structure,

3.7. the close-packed hexagonal (cph) structure,

3.8. the graphite structure, and a few indications were given on,

3.9. fibrous structures.

These considerations brought out the close relationship that exists between Nos. 3.3, 3.4 and 3.5 on the one hand and 3.7 and 3.8 on the other. Both groups of structures are closely related and are therefore subject to the same types of possible stacking faults. There exist also relations and transitions between Nos. 3.3 and 3.6. Phase transformations occur in nature between Nos. 3.3 and 3.7 (cobalt) as well as between Nos. 3.3 and 3.6 (γ-iron $\rightarrow$ α-iron). An additional type of disorder is found in graphite: the turbostratic stacking. Other examples of irregularities appear in fibrous structures.

Suggestions for Problems

As this chapter was aimed at the understanding of the translation properties of crystals, the transition between different elementary cells can be practiced. It is recommended that the metric tensor be employed. For this purpose the absolute length of vectors and the angle between vectors given by their coordinates in a crystal coordinate system may be calculated. The methods are summarised in Appendix A1. A further series of possible problems may concern the analysis of stacking faults.

4. Point Defects

4.1. General Considerations of the Field of Point Defects

Since the field of point defects is rather different from the field of dislocations we will only give a few short indications about the former here. Details can be found in the books listed in the references.

Point defects are localised in a crystal in such a way that one or several elementary cells, containing the point defect, can be virtually replaced by perfect elementary cells, resulting in an undisturbed crystal.

The elementary point defects are *vacancies* and *interstitial atoms*, and also *impurity atoms*. From these elementary defects, groups can be formed, such as double vacancies, a vacancy combined with an impurity, etc.

Point defects, in general, are mobile *configurations* having a certain energy associated with them. Apart from local changes of atomic positions the atoms can remain in place when the configuration moves. This characteristic is most evident in the case of a vacancy. When a vacancy moves, it simply exchanges places with a neighbouring atom (Fig. 4.1/1).

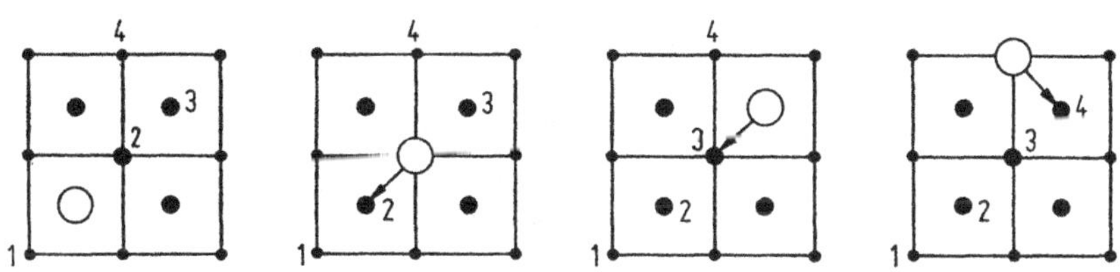

Fig. 4.1/1. Movement of a vacancy

For interstitial atoms, two possibilities exist: the single interstitial atom can either traverse the crystal on its own, or it can be repeatedly replaced on its way by other atoms, in analogy with a relay. The first case is possible, for example, in strongly anisotropic crystals such as graphite, where an atom is highly mobile between the layers but strongly bonded within them. In the face-centred cubic structure interstitial atoms, and especially impurity atoms, can be located in octahedral—(1/2 [1 1 1])—or tetrahedral—(1/4 [1 1 1]) holes, eventually leading to distortion of the surrounding lattice. On the other hand, there are arrangements such as dumbbells (Fig. 4.1/2), in which the interstitial atoms cannot be located separately, and which must obviously move along the dumbbell axis as a configuration and not as individual atoms (Fig. 4.1/3). Linear arrangements are also possible with interstitial atoms "spread" over the length of several elementary cells. These are called "crowdions". Impurity atoms in low concentration normally form individual point defects.

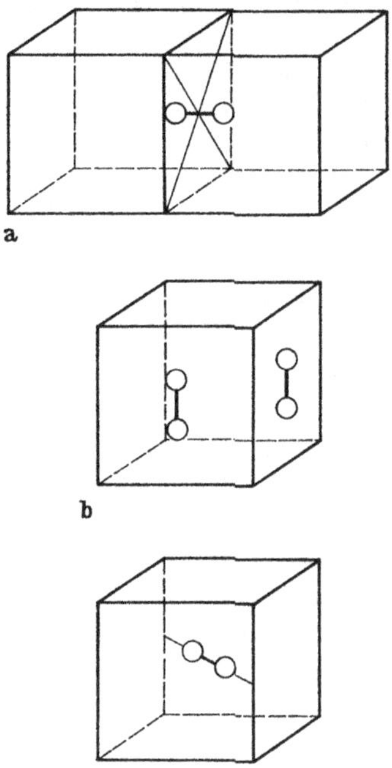

Fig. 4.1/2a–c. Dumbbell configuration of interstitial atoms a) and b) in the f.c.c. structure, c) in the b.c.c. structure

Point defects in homogeneous material can be produced spontaneously under conditions of thermal equilibrium, or by irradiation or by mechanical deformation. They can be frozen in on quenching from a state of thermal equilibrium; in this manner the material becomes supersaturated. If the material is heated so that these defects just become mobile, they will precipitate, i.e. form clusters (Fig. 10.1/1). In heterogeneous systems, point defects can be produced by alloying (sometimes involuntarily) or by the diffusion of gases into the material.

The equilibrium concentration of single point defects as a function of temperature is given by

$$c = \frac{n}{N} = e^{-\frac{E_B}{kT}} \tag{4.1-1}$$

where

c = concentration of point defects,

n = number of point defects per N atoms,

E_B = formation energy = work necessary to form the defect,

k = Boltzmann constant = $1.37 \cdot 10^{-23}$ Watt · sec · degree^{-1},

T = absolute temperature.

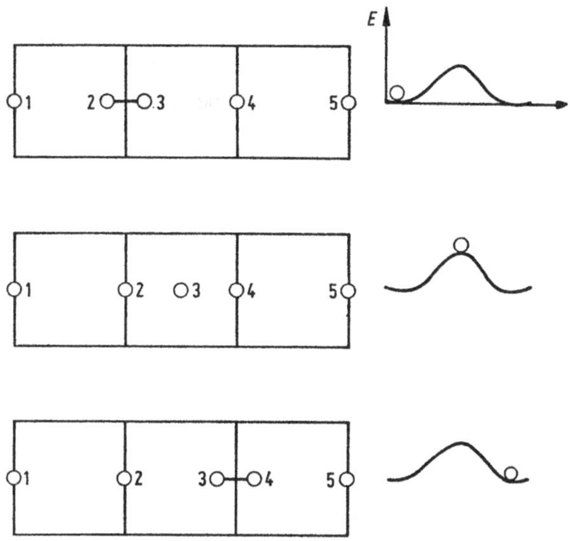

Fig. 4.1/3. Movement of a dumbbell configuration

In the Arrhenius plot, $\ln c = f(1/T)$, Eq. (4.1-1) represents a straight line, the inclination of which is determined by E_B.

The theory of the formation energy for point defects is important as it helps to identify the defect from an analysis of the thermal behaviour of the material (or the electrical resistance). A few theoretical formation energies for vacancies are:

$$\mathrm{Cu} \approx 0.8\text{–}1\ \mathrm{eV},$$

$$\mathrm{Ag} \approx 0.6\text{–}0.9\ \mathrm{eV},$$

$$\mathrm{Au} \approx 0.6\text{–}0.8\ \mathrm{eV}.$$

For the alkali metals, the energies vary from 0.55 eV for Li to 0.26 eV for Cs.

The formation energies for interstitial atoms are much higher: 3–6 eV. This means that in a state of thermal equilibrium only vacancies are present. Interstitial atoms, therefore, have to be produced by other means, e.g. by irradiation.

The formation energies of multiple point defects are, in general, smaller than the sum of the single energies; this accounts for the clustering of these defects.

If a point defect moves from one stable position to the next, it has to overcome an energy barrier. The height of this barrier thus represents the activation energy for the motion. While vacancies are formed more easily than interstitial atoms, they are less mobile. On the other hand, a double vacancy is more mobile than a single vacancy. During heat treatment of a distribution of vacancies and interstitial atoms produced by irradiation, the interstitial atoms will be the first to cluster, followed by the vacancies at a higher temperature.

Here we close our brief survey of point defects. Section 10.1 deals with the relations between point defects and dislocations.

5. The Individual Dislocation: Geometrical Basis

5.1. Outline

Dislocations are line defects in an otherwise perfect crystal. In this chapter the basic geometrical concepts shall be explained, such as the path of the line, the Burgers vector (strength of the dislocation), the relative orientation of these two, possible movements of dislocations and their relations to Moiré patterns. These latter are a convenient means for understanding many of the basic dislocation phenomena.

5.2. Definition of the Dislocation Line and the Burgers Vector

As a thought experiment one can imagine a dislocation being produced by making a cut in a perfect crystal. Next, the two sides are shifted with respect to each other by a translation vector of the lattice. Afterwards either missing material is added or surplus material is removed. The shifted surface of the cut may grow together in such a way that the cut is no longer discernible. Inside the crystal, at the edge of the former cut surface, remains a line of strongly disturbed material. This is the "dislocation line" (Figs. 5.2/1 and 5.2/2). The dislocation line is surrounded by perfect, though elastically deformed material. The cut is called a "Volterra-cut". It does not matter where this cut is made in the crystal, as long as it is limited by the given dislocation line, and as long as the displacement of the two surfaces of the cut remains the same.

Dislocations of this kind can only occur in a periodic substrate. In nature crystals are periodic substrates. In a continuum one could imagine a wedge-shaped Volterra cut with the displacement being variable along the line.

Dislocation lines are characterised by two components:

1. the path of the line in the crystal and

2. the so-called "Burgers vector", a pseudo-vector, which is a measure of the strength of the dislocation.

The material in the immediate surroundings of the dislocation line is strongly disturbed. This region is called the "dislocation core" and the material is denoted "bad material". Further away from the line the original crystalline arrangement persists, though it is elastically strained. The elastically deformed material outside the dislocation core is denoted "good material".

In order to define the Burgers vector, the disturbed crystal is compared with a perfect reference crystal. An arbitrary *line sense*, i.e. a

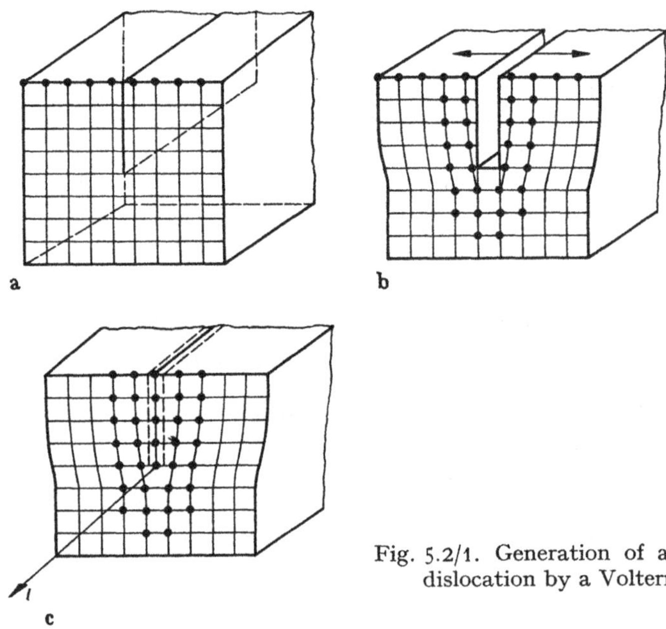

Fig. 5.2/1. Generation of an edge dislocation by a Volterra cut

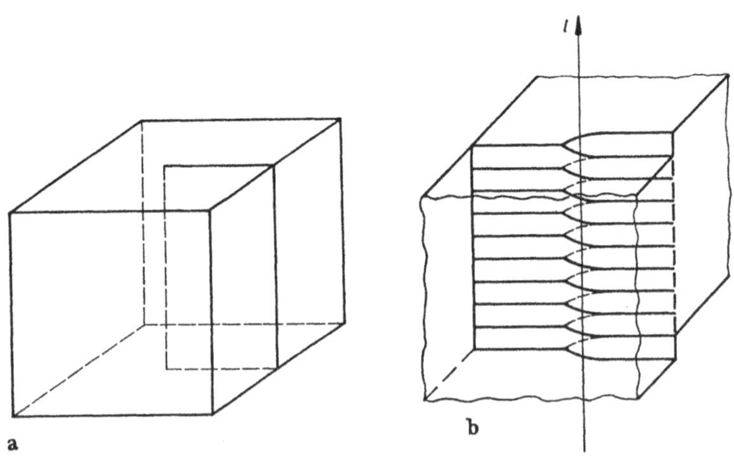

Fig. 5.2/2. Generation of a screw dislocation by a Volterra cut

direction, is attributed to the dislocation line. Now, a closed circuit in the sense of a *right-handed screw* is made around the line; it has to be situated entirely in "good" material. This closed circuit is then imaged step by step onto the reference crystal where it remains open. The vector pointing from the *end to the beginning* of the circuit *in the reference crystal* is the *Burgers vector* (Fig. 5.2/3 a).

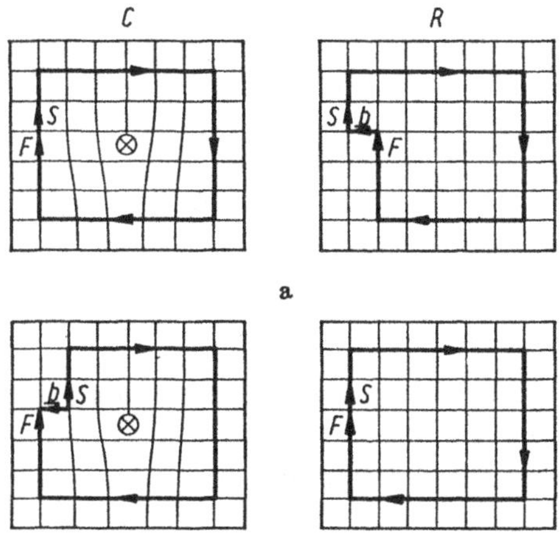

b

Fig. 5.2/3 a and b. The Burgers loops as an imaging relation between the distorted crystal C and the perfect reference lattice R. a) The loop is closed in the distorted crystal, FS/RH-convention. b) The loop is closed in the reference crystal

An equivalent procedure consists in imaging a circuit, closed in the reference crystal, onto the disturbed crystal, where it is now open. In order to maintain the sign the Burgers vector has here to point from the *beginning to the end* of the circuit (Fig. 5.2/3 b). Often the second method is more convenient for the visual representation of a dislocation. One has to follow a path in the disturbed crystal which would be closed in the perfect crystal (as many steps to the right as to the left, as many up as well as down).

Remark

In reality, we are dealing here with three different coordinate systems, i.e.

1. the Euclidean coordinate system of the surroundings,

2. the distorted system of the crystal, determined by the real positions of the atoms, and

3. the system of the perfect reference crystal.

The equivalence of the coordinate steps of systems 2 and 3 is defined by the method of imaging the circuits from one crystal onto the other. The Burgers vector can therefore be considered equally well as a *coordinate difference* in either system 2 or 3. Thus, even though the vectors connecting neighbouring atomic positions change locally in magnitude and direction with respect to system 1, due to crystal distortion, they remain constant coordinate differences, independant of location, in system 2 or 3. System 3 merely provides visual evidence of this fact. Hence, the Burgers vector has to be understood as a coordinate difference, whether the coordinate system is distorted or not.

The main points of the definition of the Burgers vector may be summarised as follows:

1. The line sense has to be arbitrarily chosen.

2. A circuit in the sense of a right-handed screw has to be made around the dislocation in such a way that

a) it is closed in the disturbed crystal, or

b) it is closed in the reference crystal.

3. a) Then, in the reference lattice, the Burgers vector points from the *end to the beginning*, and

b) in the disturbed crystal, from the *beginning to the end* of the circuit.

4. The Burgers vector is given in crystal coordinates. (The unit vectors are the edges of the elementary cell.)

This definition of the sign of the Burgers vector is given in the literature as the *FS/RH*-sign (Finish-Start/Right Hand) *in the reference lattice* (Bilby *et al.*, 1955). The opposite sign is also used (connection from the end to the beginning of the circuit in the disturbed crystal) (Read, 1953). It is important for the study of the geometry of dislocation arrangements to be aware of the sign definition applied in each particular case. Throughout this book, the FS/RH sign definition shall be used.

An enumeration of the essential properties of the Burgers vector, $\boldsymbol{b}$ follows:

a) $\boldsymbol{b}$ is independant of the path of the circuit around the dislocation line and of its starting point, as long as the dislocation is enclosed by the circuit made in the sense of a right hand screw.

b) $\boldsymbol{b}$ is independant of the position of the circuit with respect to the dislocation as long as the circuit is moved along the same dislocation, i.e. $\boldsymbol{b}$ *is invariant along the dislocation.*

A dislocation cannot start or end in the interior of a crystal, as this would mean that $\boldsymbol{b}$ would change from a finite vector to zero on moving the Burgers circuit along the dislocation. The dislocation either starts or ends at a surface, or forms a closed loop.

c) The sign of $\boldsymbol{b}$ has significance only when considered together with the line sense. $\boldsymbol{l}$ is a vector in the dislocation line indicating the chosen line sense and the local line orientation. If $\boldsymbol{l}$ and $\boldsymbol{b}$ change their signs simultaneously, they still represent the same unchanged geometrical situation in the crystal. This is due to the fact that, from the definition of $\boldsymbol{b}$, the Burgers circuit has to be made around the line in the sense of a right-handed screw. If the line sense changes, the sense of the circuit also changes, and with it the sign of the vector connecting the beginning to the end of the circuit in the disturbed crystal. Every dislocation is therefore characterized by two equivalent descriptions: $(\boldsymbol{b}, \boldsymbol{l})$ and $(-\boldsymbol{b}, -\boldsymbol{l})$. On the other hand, $(\boldsymbol{b}, -\boldsymbol{l})$ and $(-\boldsymbol{b}, \boldsymbol{l})$ respecitvely describe a different dislocation. A geometrical function, $F(\boldsymbol{b}, \boldsymbol{l})$ of dislocations has to be formulated in such a way that

1. $F(\boldsymbol{b}, \boldsymbol{l}) \equiv F(-\boldsymbol{b}, -\boldsymbol{l})$, and (5.2-1)

2. $F(\boldsymbol{b}, \boldsymbol{l}) \neq F(\boldsymbol{b}, -\boldsymbol{l})$ or $F(-\boldsymbol{b}, \boldsymbol{l})$, respectively. (5.2-2)

The following distinction has to be made between a vector and a pseudo-vector: A vector is fixed in space, independant of the coordinate system; e.g. if the coordinate axes (unit vectors) are inverted $\boldsymbol{u}_i \rightarrow -\boldsymbol{u}_i$ $(i=1, 2, 3)$, the coordinates also change their sign $\xi^i \rightarrow -\xi^i$; a pseudo-vector, however, changes sense along with the coordinate system so that the values of the coordinates remain unchanged, i.e. $\xi^i \rightarrow \xi^i$. $\boldsymbol{b}$ is therefore a pseudo-vector with $\boldsymbol{l}$ fixing a coordinate axis.

We have also to distinguish clearly between the two terms "line sense" and "line orientation". A straight line can, for example, lie in a North-South orientation, and can point in the "sense" of the arrow towards either North or South. In every orientation, therefore, there exist two possibilities for the line sense.

d) $\boldsymbol{b}$ has to be a translation vector of the lattice because only then can the crystal be "perfect" although elastically distorted (good material) outside the dislocation line. If $\boldsymbol{b}$ were *not* a translation vector, the crystal would be disturbed over the whole surface of the Volterra cut. This will become evident in Section 8.1 which deals with partial dislocations.

5.3. Relative Orientation of $\boldsymbol{b}$ and $\boldsymbol{l}$

The Burgers vector is constant in length as well as in orientation and sense along a dislocation line. The dislocation line, however, can arbitrarily change its direction. Thus, the angle between $\boldsymbol{b}$ and $\boldsymbol{l}$ can take any value, but the following two limiting cases are important:

1. $\boldsymbol{b} \perp \boldsymbol{l}$ is termed "edge orientation", and
2. $\boldsymbol{b} \parallel \boldsymbol{l}$ the "screw orientation".

Straight dislocations in these limiting orientations are often called edge or screw dislocations. In the edge orientation an additional plane appears to be built into the crystal (Fig. 5.2/1). The term "screw orientation" expresses the fact that the original parallel crystal planes are now joined as a helical surface (Fig. 5.2/2). Fig. 5.3/1 shows a dislocation changing from edge to screw orientation.

A general dislocation can be understood as being formed of a screw and an edge component. The Burgers vectors of these components are:

a) screw component $b_{\text{screw}} = b \cdot \cos(\alpha)$, (5.3-1)

b) edge component $b_{\text{edge}} = b \cdot \sin(\alpha)$, (5.3-2)

where α is the angle between $\boldsymbol{l}$ and $\boldsymbol{b}$. ($\alpha = 0 =$ screw orientation, $\alpha = \pi/2 =$ edge orientation.)

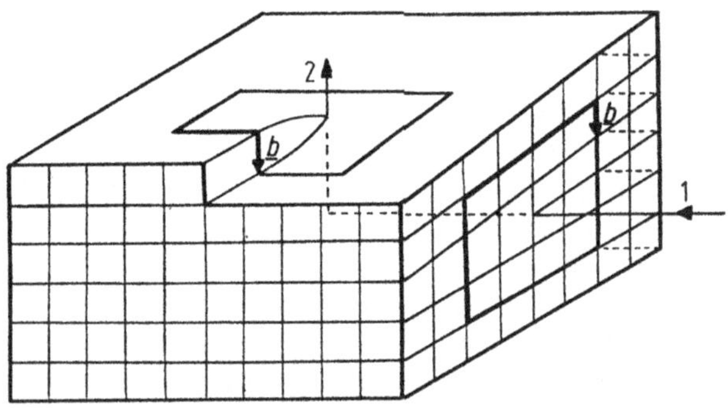

Fig. 5.3/1. Change of a dislocation from the edge- (*1*) to the screw-orientation (*2*)($\boldsymbol{b}$ is drawa here in the SF/RH convention!)

By the sign definition (FS/RH), used in this book, a right-handed screw is represented by $\boldsymbol{b} \upuparrows \boldsymbol{l}$, and a left-handed screw by $\boldsymbol{b} \uparrow\downarrow \boldsymbol{l}$. The character of a screw component can be determined in the following way:

$$\operatorname{sgn}(\boldsymbol{l} \cdot \boldsymbol{b}) = \begin{cases} + \text{ right-handed screw} \\ - \text{ left-handed screw} \end{cases} \qquad (5.3\text{-}3)$$

Using the FS/RH convention, the right handed screw becomes positive. This harmonises better than the negative sign with our predominantly right-handed world.

It might be of interest to determine from a given $\boldsymbol{l}$ and $\boldsymbol{b}$ the position of the additional plane of the edge component in a general dislocation. Instead of going through the definition of the Burgers vector one may write

$$\boldsymbol{p} = [\boldsymbol{l} \times \boldsymbol{b}]. \qquad (5.3\text{-}4)$$

As the crystal is virtually divided into two parts by a plane which is determined by $\boldsymbol{l}$ and $\boldsymbol{b}$, the vector $\boldsymbol{p}$ points into the part containing the additional layer. The two geometrical statements (5.3-3) and (5.3-4) fulfill the invariance condition: $F(\boldsymbol{l}, \boldsymbol{b}) \equiv F(-\boldsymbol{l}, -\boldsymbol{b})$ [see (5.2-1)].

5.4. Motion of Dislocations

One may draw an analogy between a dislocation and a water wave: A wave can pass along while the water particles move in circles around one spot. A dislocation moves in a similar fashion, whereby individual atoms are only slightly displaced. The motion of dislocations is the actual process of plastic deformation. A dislocation line can, in principle, move arbitrarily (except along its own extension). However, two limiting cases must again be distinguished:

1. *Glide*, the motion in the glide plane which is determined by the path of the dislocation line and by the direction of the Burgers vector. This motion involves no mass transport, i.e. no diffusion. The atoms above and below the glide plane are displaced by $\boldsymbol{b}$ with respect to each other (Fig. 5.4/1). Since, in the screw orientation, the directions of $\boldsymbol{b}$

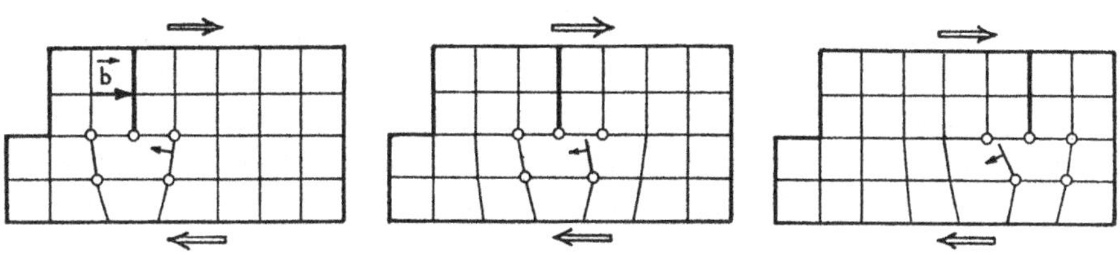

Fig. 5.4/1. Motion of a dislocation by glide

and $\boldsymbol{l}$ coincide, a screw dislocation can, in principle, glide on any plane passing through the dislocation line.

2. *Climb*. In pure climb the dislocation moves perpendicularly to the glide plane. This motion is only possible if

a) the orientation of the dislocation has an edge component, and

b) if the temperature or the stress are sufficiently high so that diffusion becomes possible. Fig. 5.4/2 shows that climb involves the build up or removal of the additional plane.

At low temperature glide is the only mechanism responsbile for the plastic deformation, whereas climb enters into part at high temperature processes (recrystallisation, creep).

If a dislocation moves it divides the crystal virtually into two parts, one of which is displaced with respect to the other. The amount and

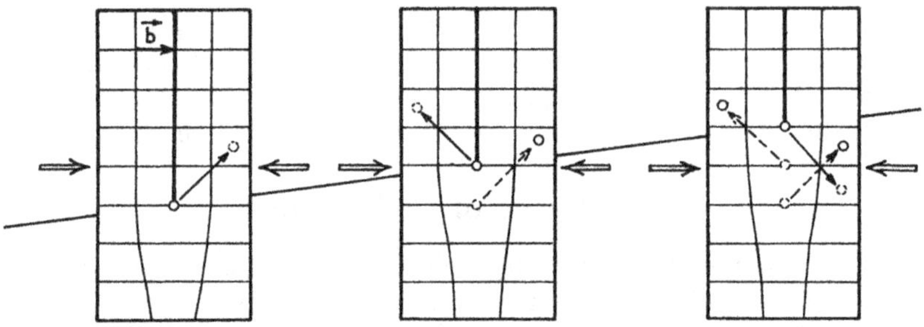

Fig. 5.4/2. Motion of a dislocation by climb

direction of this displacement can be determined by the following rule, which we term the "movement rule". First the following vector product is formed:

$$[\boldsymbol{l} \times \boldsymbol{v}] = \boldsymbol{m} \tag{5.4-1}$$

where

$\boldsymbol{l}$ is the direction of the line, and

$\boldsymbol{v}$ is the direction of the motion ($\boldsymbol{v}$ is reminiscent of the velocity vector).

The vector $\boldsymbol{m}$, as defined by Eq. (5.4-1), thus is normal to the plane swept over by the dislocation and **m**arks that part which is displaced by $+\boldsymbol{b}$ while the other part remains fixed. This rule holds for any motion of dislocations in either edge or screw orientation, for glide as well as for climb. It fulfills the invariance condition (5.2-1). Fig. 5.4/3 shows two examples; one for glide and the other for climb.

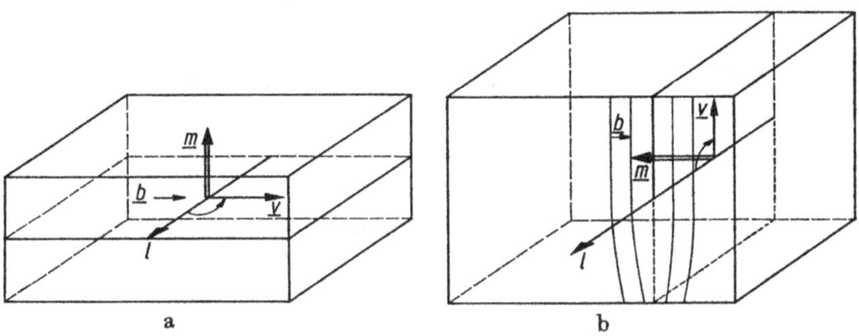

Fig. 5.4/3a and b. Movement rule. a) Glide, b) climb

The movement rule offers the possibility of defining the Burgers vector without taking into account the distortion of the crystal. One only has to compare the states of the crystal before and after the passage of the dislocation. The crystal is perfect and undistorted in both cases. The relative displacement of the two parts of the crystal between these states related to the vectors $\boldsymbol{l}$, $\boldsymbol{v}$ and $\boldsymbol{m}$, according to Eq. (5.4-1), is the Burgers vector.

5.5. Dislocations and Moiré-Figures

Let us assume two model crystals consisting of homogenious layers, vertical to the plane of the paper (Fig. 5.5/1a); the crystal 1 lies behind that plane and the crystal 2 in front of it. Initially, there shall be no

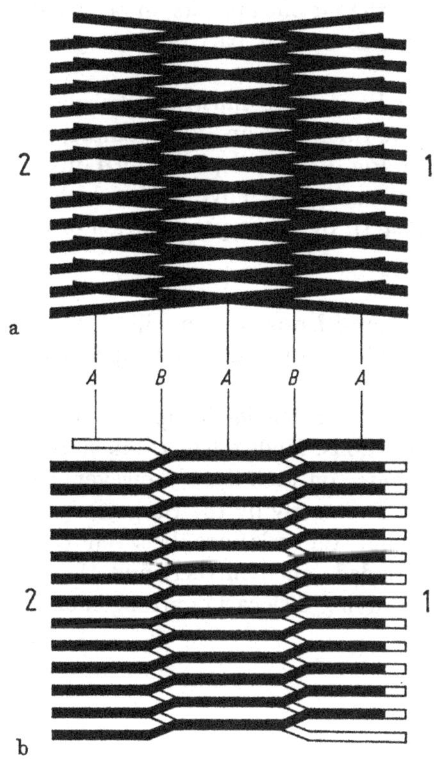

Fig. 5.5/1. a) Moiré pattern, b) corresponding screw dislocations

interaction between them. There are areas, in which the crystals match perfectly, that is where the lines cross (A), and areas of particularly bad matching, i.e. where the two sets of lines are separated by half the line spacing (B). This situation can be represented by the moiré model No. 4. We obtain a series of light and dark bands. The light bands are areas of good fit and the dark bands of bad fit.

Now we assume the two crystals to grow together. The regions of good fit (A) will extend by relaxation and those of bad fit (B) will contract. Since the crystal prefers order to disorder, the bands of bad fit will be contracted as far as possible; these bands are the dislocations (Fig. 5.5/1 b). The dislocations thus lie in the middle of the dark moiré bands. In this way, dislocations (screw dislocations in our example) can be represented by means of rigid moiré models.

Edge dislocations can be represented by two series of lines in the same orientation, but with slightly different spacings. This situation can also be shown with model No. 4, only here the two sheets have to be held at a distance so that the far one appears, in perspective, to be shortened. This region corresponds to that part of the crystal containing the additional plane.

The models can be used to illustrate the glide mechanism and to check the movement rule. We can, for example, assume that the front part of the crystal in Fig. 5.5/1 is displaced by $\boldsymbol{b}$ (upwards) and determine to which side the dislocation will move, i.e. in which direction the vector $\boldsymbol{v}$ will point. From (5.4-1), by cyclic permutation of the terms, we obtain:

$$\boldsymbol{v} = [\boldsymbol{m} \times \boldsymbol{l}] \,. \tag{5.5-1}$$

If $\boldsymbol{m}$ points forward and $\boldsymbol{l}$ upwards, $\boldsymbol{v}$ will point to the left, i.e. if the front part ($\boldsymbol{m}$) is displaced in direction $\boldsymbol{b}$ upwards, the dislocation will move to the left.

A twist of the Moiré models in the opposite direction represents left-handed screw dislocations. Models No. 4, taken as edge dislocation, can be used for analogous considerations; however, the signs for $\boldsymbol{l}$ and $\boldsymbol{b}$ have first to be fixed correctly by the Burgers circuit of by (5.3-3).

We shall see that the relations between dislocations and Moiré figures can be extended much further, and that the latter provide a useful means for studying dislocation arrangements.

6. Interaction between a Dislocation and its Surroundings

6.1. Outline

Up to now the situation and behaviour of a dislocation have been considered essentially independently of the surroundings. Now, these surroundings appear in the form of an external stress field acting on the dislocation. The forces which can move the dislocation originate from this stress field. In this chapter a representation of the stress field will first be given and afterwards the forces on a dislocation will be discussed.

Dislocations have associated with them a certain amount of energy, i.e. the work required for their formation. A curved dislocation line can lower its energy by stretching, hence it is subject to a line tension. Furthermore, the dislocation itself produces strain and stress fields which, in turn, act on other dislocations, as will be shown in the following chapter.

6.2. Representation of the Stress Field

We introduce here the necessary basic notions and refer to the literature for the further development of the theory of elasticity, e.g. Hirth and Lothe (1968).

Different types of forces can act on a body, or specifically on a crystal: forces originating at the external surface, for example at the grips in a tensile specimen, or forces spread over the whole crystal, such as gravitation or magnetic forces (volume forces). Internal forces can also operate, e.g. forces due to inhomogeneous thermal expansion. We consider a static crystal, i.e. a crystal in which all the forces acting are distributed in such a way that their sum, as well as the sum of the angular momenta, is zero.

In order to determine the forces acting inside the crystal at a given point we may imagine an infinitesimal cube cut out of the surroundings of the point under investigation. Next we assume that forces are homogeneously distributed over the cube faces such that the cube is placed into the same state of stress as that which existed inside the crystal. Since the volume forces decrease with the third power of the linear dimensions they can be neglected. The force per unit surface area is called stress. $\boldsymbol{\sigma}_1$ is the stress on that cube face on which the positive x-axis is perpendicular and pointing outwards (accordingly, $\boldsymbol{\sigma}_2$ and $\boldsymbol{\sigma}_3$ for the positive y- and z-axis). $\boldsymbol{\sigma}_{-1}$ is the stress on the opposite side of the cube. Each of these stresses is split into three components corresponding to the three directions of the coordinate axes: e.g. $(\boldsymbol{\sigma}_1 = \sigma_{11}, \sigma_{12}, \sigma_{13})$ (Fig. 6.2/1).

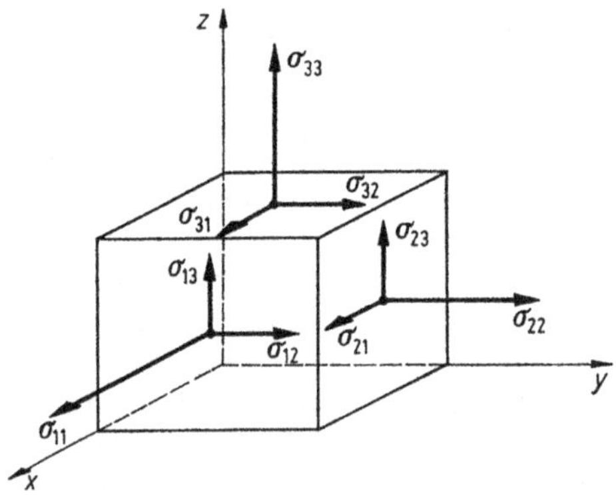

Fig. 6.2/1. Stress components on the surface of an "infinitesimal" cube

From the condition that the cube should be immobile the following relation holds for the equilibrium of forces:

$$\sigma_{-i-k} = -\sigma_{ik} \qquad (i = 1 \ldots 3,\ k = 1 \ldots 3) \tag{6.2-1}$$

and, from the equilibrium of the angular momenta, we have

$$\sigma_{ki} = \sigma_{ik}. \tag{6.2-2}$$

This leads to 6 independent components for the stresses, forming a symmetrical tensor:

$$\begin{pmatrix} \sigma_{11} & \sigma_{12} & \sigma_{13} \\ \sigma_{21} & \sigma_{22} & \sigma_{23} \\ \sigma_{31} & \sigma_{32} & \sigma_{33} \end{pmatrix}. \tag{6.2-3}$$

The diagonal elements, σ_{ii}, if they are positive, stand for *tensile stresses*, otherwise for *compressive stresses*. Components with mixed indices σ_{ik} are *shear stresses*. Occasionally, the shear stresses are indicated by τ. This tensor defines all the stresses at one point (except for the case where internal torques have to be taken into account (see Hirth and Lothe, 1968, p. 31). Obviously all six components of the stress tensor may depend on the location within the crystal. Elasticity theory deals with all the possible stress functions and the relations between stress and strain. We shall apply the results obtained by elasticity theory; for their derivation we refer again to Hirth and Lothe (1968).

6.3. Forces Acting on a Dislocation

In order to analyse the effect of "forces" on a dislocation we have to remember that a dislocation is moving through a crystal as a configuration which is not linked to specific particles and therefore does not present a fixed point for the forces to act on.

If, on the other hand, we admit that the dislocation configuration possesses a certain physical individuality, we can also assume that its motion is caused by "forces". In this case we can derive the "force" from the definition of the "work" (= force × distance moved), i.e. force = work per unit distance moved.

Fig. 6.3/1 shows a crystal traversed by an edge dislocation, with a shear stress, τ, acting on the surface of the crystal. If the dislocation moves through the whole crystal, and thus virtually divides it into two parts, one part is displaced with respect to the other by $\boldsymbol{b}$ [see the movement rule Eq. (5.4-1)]. The work done by the total force on the crystal is then given by:

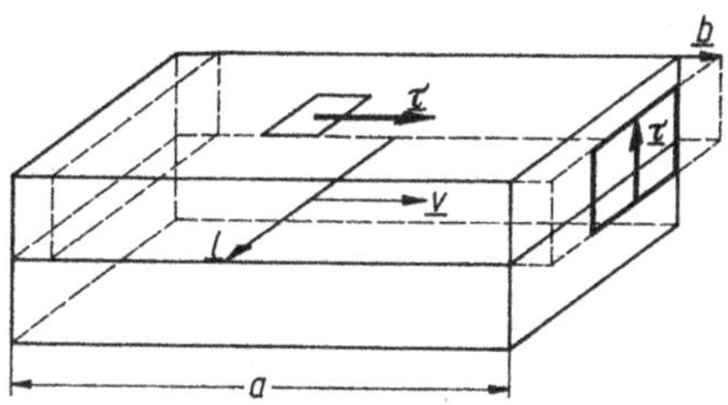

Fig. 6.3/1. Work done on a crystal by a dislocation

Work = Shear stress (force per unit surface) × surface × distance $|\boldsymbol{b}|$

$$W = \tau (l \cdot a) \cdot b. \tag{6.3-1}$$

This work can also be considered as the *work done by a force on the dislocation*:

Work = force per unit length of the dislocation
× length of the dislocation
× distance moved by the dislocation

$$W = f \cdot l \cdot a. \tag{6.3-2}$$

From the comparison of these two equations it follows that the force per unit length of the dislocation is given by:

$$f = \tau \cdot b. \tag{6.3-3}$$

This force, f, is thus proportional to the shear stress τ and the absolute value of the Burgers vector, b.

In general, the differential of the force, $\boldsymbol{df}$, on a dislocation is given by the formula of Peach and Koehler*:

$$\boldsymbol{df} = |\boldsymbol{b}|\,[\boldsymbol{\sigma}_b \times \boldsymbol{dl}] \tag{6.3-4}$$

with

$\boldsymbol{b}$ = Burgers vector,

$\boldsymbol{dl}$ = vector element in the direction of the dislocation line,

$\boldsymbol{\sigma}_b$ = stress on the plane normal to the Burgers vector, and on which $\boldsymbol{b}$ points outwards.

The formula indicates that $\boldsymbol{df}$ is always normal to $\boldsymbol{dl}$, i.e. the force always acts normal to the dislocation line. It has no tangential component in the line. The dislocation line cannot move within itself; forces in the direction of the line are therefore meaningless. Furthermore, $\boldsymbol{df}$ is also normal to $\boldsymbol{\sigma}_b$. Fig. 6.3/2 shows an example of a force with a component in the glide plane. In contrast to the example of Fig. 6.3/1 the vertical component of the shear stress has to be taken into account. However, due to the symmetry of the stress tensor both components are equal.

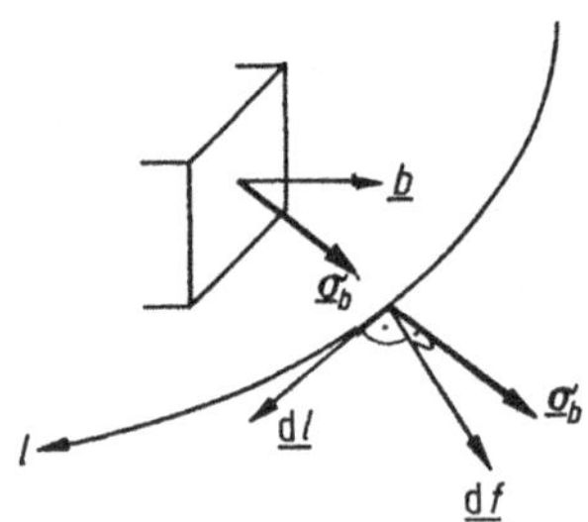

Fig. 6.3/2. Peach and Koehler formula

The Peach and Koehler formula also applies to climb but, as shown by Weertman (1964), the conditions have to be defined more precisely. Under hydrostatic pressure, P, the dislocations do not climb, even if the temperature is sufficiently high to provide the necessary mobility of the vacancies. It is therefore meaningful to replace the general stress, σ_{ik} by $\sigma_{ik}+P$ with $P=-1/3\;(\sigma_{11}+\sigma_{22}+\sigma_{33})$ = hydrostatic pressure. The external stress, defined by the Peach and Koehler formula, may be balanced by other forces, such as "friction" or "chemical" forces, the latter being produced, for example, by the change in the equilibrium concentration of vacancies on movement of the dislocation.

* Privately communicated in this form by Professor J. S. Koehler.

6.4. Energy of a Dislocation

Work is required for the creation of a dislocation in a crystal. This work is stored inside the crystal as positive potential energy. The energy of a dislocation, naturally, depends on its length but, for dislocations of equal length it is further determined by their screw or edge character as well as by their curvature. In the following we calculate the energy of a length L of an infinitely long straight dislocation in a screw orientation.

Fig. 6.4/1 shows an unrolled cylindrical element of a screw dislocation. This element is deformed in shear from a rectangle into a parallelogram. The angle γ is given by:

$$\gamma = \frac{b}{2\pi r} \tag{6.4-1}$$

with

γ = shear strain,

b = length of the Burgers vector = displacement,

r = radius of the cylinder.

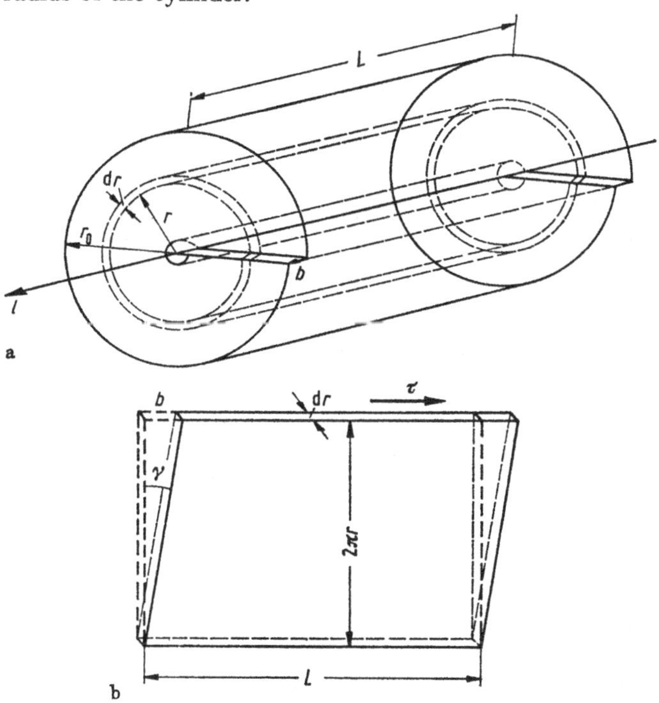

Fig. 6.4/1 a and b. Geometrical situation for the calculation of the energy of a screw dislocation. a) Volume element inside the cylinder, b) evolution of the volume element

The required shear stress, τ is defined by

$$\tau = G\gamma \tag{6.4-2}$$

G = shear modulus (for iron $\approx 8 \cdot 10^{11}$ dynes/cm^2 $\approx$ 8,000 kg/mm^2).

By substituting the relation for the shear strain, γ we obtain

$$\tau = \frac{Gb}{2\pi r}. \tag{6.4-3}$$

The elastic energy per unit volume, u is

$$u = \tfrac{1}{2}\tau\gamma. \tag{6.4-4}$$

This is equal to the work done during the elastic deformation of a volume element. Substitution of the above values for τ and γ gives:

$$u = \frac{1}{8}\,\frac{Gb^2}{\pi^2 r^2}. \tag{6.4-5}$$

The volume element is

$$dV = L \cdot 2\pi r \cdot dr. \tag{6.4-6}$$

Thus the energy content of this cylindrical volume element becomes:

$$dU = u \cdot dV = \frac{Gb^2 L}{4\pi} \cdot \frac{dr}{r}. \tag{6.4-7}$$

In order to calculate the total energy of the dislocation we have to split it into two parts:

1. the core, i.e. the "bad" material, in which the deformation is so large that the laws of elasticity are no longer valid. We assume the radius of this core to be b, i.e. the Burgers vector. This corresponds very roughly to one atomic spacing,

2. the elastically deformed "good" material.

The total energy follows as:

$$U = U_c + \frac{Gb^2 L}{4\pi}\int_b^{r_0} \frac{dr}{r} = \Gamma_c L + \frac{Gb^2 L}{4\pi}\ln\left(\frac{r_0}{b}\right), \tag{6.4-8}$$

where U_c is the core energy.

Discussion

1. $U \propto L$. This holds, in principle, for long straight parts of the dislocation. We can therefore write: $U = \Gamma \cdot L$, with Γ being the energy per unit length.

2. Γ contains the term Γ_c, the energy of the dislocation core per unit length. We obtain a rough estimate of this energy by considering the dislocation core, with its highly disturbed atomic arrangement, as a

row of atoms in the liquid state. In this case Γ_c equals the number of atoms per unit length times the heat of fusion per atom. A closer estimation is obtained from:

$$\Gamma_c = 2\frac{G b^2}{4\pi}. \tag{6.4-9}$$

This leads to

$$\Gamma = \frac{G b^2}{4\pi}\left(\ln\frac{r_0}{b} + 2\right). \tag{6.4-10}$$

3. Γ increases with $\ln(r_0/b)$, i.e. if r_0 tends to infinity, Γ also becomes infinitely large, but slower than r_0. This means that in an infinitely large crystal the value of the energy per unit length of a dislocation also becomes infinite. However, as no infinitely large crystals exist, r_0 has to be arbitrarily fixed. For example, r_0 can be the size of the crystal, e.g. 1 cm, or the size of a crystal grain, e.g. 10 μ, or in highly deformed material the distance to the next dislocation. If we put $r_0 = 70\,b \approx 170\,\text{Å}$, Γ becomes

$$\Gamma = \frac{G b^2}{2} \tag{6.4-11}$$

and for $r_0 = 4\cdot 10^4\, b \approx 10\,\mu$:

$$\Gamma = G b^2. \tag{6.4-12}$$

4. $\Gamma \propto$ shear modulus G. This means that in a "hard" crystal (hard with respect to elastic deformation) the energy of the dislocation is higher than in a soft one. This is obvious, since Γ is the work necessary to produce a unit length of dislocation.

5. $\Gamma \propto b^2$ can be explained by the fact that the elastic force is proportional to the displacement, i.e. $\propto b$, and that the energy (work = force × distance moved) is thus proportional to b^2; this proportionality is useful in the estimation of dislocation interactions. On the other hand, we can also see immediately that dislocations with the smallest possible Burgers vector will be preferentially formed, since for two distant dislocations the energy is $\propto b^2 + b^2 = 2b^2$, whereas for united dislocations it is $\propto (2b)^2 = 4b^2$.

6. Up to now we have only dealt with screw dislocations. The energy of an edge dislocation is higher:

$$\Gamma_{\text{edge}} = \frac{\Gamma_{\text{screw}}}{1-\nu}. \tag{6.4-13}$$

ν = Poisson's ratio $= -(\Delta w/w)/(\Delta l/l) \approx 1/3$,

$(\Delta w/w)$ = relative change of width (perpendicular to the traction),

$(\Delta l/l)$ = relative change of length (direction of traction),

w = width (diameter of rod),

l = length.

Since ν is $\approx 1/3$ in usual material, we have

$$\Gamma_{\text{edge}} = 3/2\,\Gamma_{\text{screw}}. \tag{6.4-14}$$

The energy of a dislocation in edge orientation is thus about 50% higher than that of a dislocation in screw orientation.

7. The energy of a dislocation of arbitrary character can be calculated by summing the energies of the screw and edge components. If α is the angle between $\boldsymbol{b}$ and $\boldsymbol{l}$, the screw component of the Burgers vector is $b \cdot \cos\alpha$ and the edge component $b \cdot \sin\alpha$. The energy per unit length is then:

$$\Gamma = \Gamma_0\left(\cos^2\alpha + \frac{\sin^2\alpha}{1-\nu}\right) \tag{6.4-15}$$

with

$$\Gamma_0 = \frac{G b^2}{2}. \tag{6.4-16}$$

As an example, we may estimate the energy stored by dislocations in 1 cm³ of heavily deformed iron ($G = 8.2 \cdot 10^3$ kg/mm², $b = 2.5$ Å). We assume a dislocation density, D of 10^{12} dislocations per cm². The energy is given by:

$$\text{energy} = D\,\frac{G b^2}{2} \approx 2.4\ \text{kg} \cdot \text{m}. \tag{6.4-17}$$

If this energy were transformed spontaneously into heat the cubic centimeter of iron would warm up by about 7 °C.

Some numerical indications:

The shear modulus, G can be expressed in terms of the elastic modulus, E and the Poisson's ratio, ν by:

$$G = \frac{E}{2(1+\nu)} \approx \frac{E}{2.6}. \tag{6.4-18}$$

Metal	E (dynes/cm²)	G (dynes/cm²)	a (A)	$b_{\min}$ (A)
Al	$7 \cdot 10^{11}$	$2.6 \cdot 10^{11}$	4.04	2.86
Ni	$20 \cdot 10^{11}$	$7.5 \cdot 10^{11}$	3.52	2.52
Cu	$12 \cdot 10^{11}$	$4.5 \cdot 10^{11}$	3.60	2.54
Au	$8 \cdot 10^{11}$	$3.0 \cdot 10^{11}$	4.07	2.88
Fe	$20 \cdot 10^{11}$	$8.2 \cdot 10^{11}$	2.86	2.47

10^{11} dynes/cm $\approx 1{,}000$ kg/mm² (gravity $g = 9.81$ m/sec² ≈ 10 m/sec²).

6.5. The Line Tension

Since the energy of a dislocation is proportional to its length, the dislocation line will tend to adjust itself, analogous to a rubber band, in such a way that its length—and therefore also its energy—becomes as small as possible. Hence a force, the so-called line tension, T, is acting along the line.

$$T = \frac{dU}{dL} = \Gamma = \frac{Gb^2}{2}. \tag{6.5-1}$$

The extension from Eq. (6.4-9) to (6.4-10) and (6.5-1), respectively, is a rough estimation. The energy U of a dislocation depends on the character of the dislocation, i.e. edge—or screw orientation, as well as on its curvature and surroundings. A detailed discussion can be found in Hirth and Lothe (1968), p. 150.

A special problem was treated by Brown (1964) (see also Hirth and Lothe, 1968, p. 151). If a *screw dislocation* bows out it becomes longer and develops edge components. The bowing out and the formation of edge components both contribute to an increase in energy. On the other hand, an *edge dislocation* also becomes longer on bowing out, thus increasing its energy; however, as in this case screw components with lower specific energy are formed the increase in energy due to the bowing out is partly compensated by the decrease on the formation of screw components. A screw dislocation is therefore more resistant to bowing out than an edge dislocation.

As an example, we consider a circular dislocation, with radius r, in its glide plane. We assume the dislocation to be of uniform character, and thus disregard the edge- and screw orientation. If no forces were acting on the dislocation it would contract and disappear. According to the Peach and Koehler formula (6.3-4) a shear stress τ acts as a force, $f = \tau \cdot b$, on a unit length of the dislocation. f is normal to the dislocation line. We must now determine what shear stress τ would maintain in equilibrium the dislocation loop of radius r. We assume that the radius increases from r to $r + dr$, with the force τb providing the work necessary for this increase. This work can be understood as the work done by the line tension in increasing the length of the dislocation (Fig. 6.5/1), i.e.:

Work = force per unit length × length × distance moved
= line tension × increase in length of the line

$$dW = \tau b \cdot 2\pi r \cdot dr = \frac{Gb^2}{2} \cdot 2\pi\, dr \tag{6.5.-2}$$

and thus

$$\tau = \frac{Gb}{2r}. \tag{6.5-3}$$

The shear stress τ, necessary to maintain a circular dislocation with radius r, is proportional to $1/r$, i.e. the larger the loop, the smaller the required shear stress. The radius r can also be considered as a local radius of curvature, so that Eq. (6.5-3) can also be applied to dislocation segments.

The deformation of a metal is at first essentially elastic. Only when the tension exceeds a certain limit, the yield point, does plastic deformation start. Some values of yield points are 15–85 kg/mm² for polycristalline carbon steels and about 1 kg/mm² for pure metal single

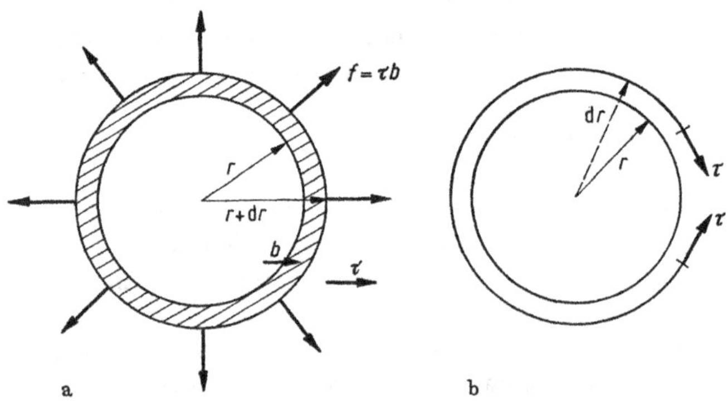

Fig. 6.5/1 a and b. Extension of a dislocation loop under the action of a shear stress τ. a) Work done by the force $\boldsymbol{f}$ on the dislocation, b) work of the line tension T (not τ as erroneoulsy indicated in the drawing)

crystals. On the basis of the above considerations we shall try to get an understanding of these values.

We assume that a straight dislocation is moved forward by a shear stress. In its path it is locked at two obstacles, e.g. precipitates or other dislocations, spaced by the distance L. We now calculate the shear stress, τ necessary to force the dislocation through the gap between the obstacles.

At first the dislocation is straight ($r=\infty$). Then it bows out (r decreases as τ increases). As long as the curvature decreases with increasing τ, the arrangement is stable. The radius of curvature reaches a minimum at $r=L/2$. With further increase of τ the situation becomes unstable, as r starts to increase again. The loop expands spontaneously. For iron ($G=8\cdot 10^3$ kg/mm², $b=2.8$ Å and $L=1$ μ) $\tau_{\max}$ is:

$$\tau_{\max}=\frac{Gb}{L}\approx 2.2\ \text{kg/mm}^2 .$$

With these considerations we already fall within an order of magnitude of the yield strength σ_y for single crystals.

If only one single dislocation loop were expanding, the material would glide by the amount of one Burgers vector. This could not be detected macroscopically. In order to enable glide over longer distances many dislocations of the same kind must be produced. This is done by the *Frank-Read mechanism.*

We assume that a part of a dislocation is anchored at two points [e.g. by precipitates or in a dislocation network (Section 11.2)]. Under increasing shear stress it becomes curved to the shape of a semicircle. (We neglect here again the difference between screw- and edge orientation as well as the crystal anisotropy.) From now on the dislocation can extend spontaneously under constant shear stress, as shown in Fig. 6.5/2,

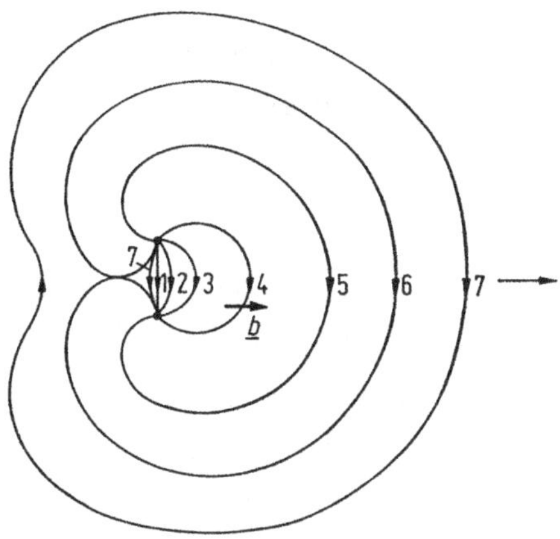

Fig. 6.5/2. Frank-Read source. (The numbering refers to the time sequence)

until finally a closed loop and a separate dislocation between the obstacles are formed. With the latter the initial configuration is reproduced. Further dislocation loops can thus be formed by the same mechanism, causing a glide displacement of $\boldsymbol{b}$. The closing of the dislocation loop, i.e. the transition from No. 6 to No. 7 in Fig. 6.5/2 is a dislocation reaction. Those will be discussed in Section 7.4.

6.6. The Strain Field of a Dislocation

Starting from the stationary position $\boldsymbol{x}_0$ in a perfect crystal the strain is described as a deviation from the initial position by an additional vector $\boldsymbol{u}$ (not to be confused with the unit vector). Thus, the real position of the atoms is given by:

$$\boldsymbol{x} = \boldsymbol{x}_0 + \boldsymbol{u}(\boldsymbol{x}_0) . \tag{6.6-1}$$

Every dislocation is surrounded by a strain field which is described by the space function of the displacement vector $\boldsymbol{u}$ with the components (u, v, w).

We indicate here the results for a dislocation in screw- and one in edge orientation (Read, 1963, p. 115). The dislocation line shall point in the positive z-direction and the material is assumed to be elastically isotropic.

Strain Field of a Screw Dislocation (right-handed screw)

in Cartesian coordinates (x, y, z):

$$\begin{aligned} u(x, y, z) &= 0 \\ v(x, y, z) &= 0 \\ w(x, y, z) &= \frac{b}{2\pi} \arctan \frac{y}{x}; \end{aligned} \tag{6.6-2}$$

in cylindrical coordinates (r, θ, z):

$$\begin{aligned} u(r, \theta, z) &= v(r, \theta, z) = 0 \\ w(r, \theta, z) &= \frac{b}{2\pi} \theta. \end{aligned} \tag{6.6-3}$$

It is obvious that the displacement, which transforms the xy-plane into a screw plane, points in the z-direction (Fig. 6.4/1).

Strain Field of an Edge Dislocation

in Cartesian coordinates (x, y, z):

(The dislocation line is assumed to lie in the z-axis and the extra half-plane in the positive y-axis.)

$$\begin{aligned} u(x, y, z) &= \frac{b}{2\pi} \left[\arctan\left(\frac{y}{x}\right) + \frac{1}{2(1-\nu)} \frac{xy}{x^2+y^2} \right] \\ v(x, y, z) &= \frac{-b}{8\pi(1-\nu)} \left[(1-2\nu) \cdot \ln(x^2+y^2) + \frac{x^2-y^2}{x^2+y^2} \right] \\ w(x, y, z) &= 0 \end{aligned} \tag{6.6-4}$$

$\nu =$ Poisson's ratio (Section 6.4).

Fig. 6.6/1 represents a calculated example of this strain field ($\nu = 1/3$);

in cylindrical coordinates (r, θ, z):

(With the angle θ starting on the positive x-axis.)

$$\begin{aligned} u(r, \theta, z) &= \frac{b}{2\pi} \left[\theta + \frac{\sin(2\theta)}{4(1-\nu)} \right] \\ v(r, \theta, z) &= -\frac{b}{2\pi} \left[\frac{1-2\nu}{2(1-\nu)} \ln(r) + \frac{\cos(2\theta)}{4(1-\nu)} \right] \\ w(r, \theta, z) &= 0. \end{aligned} \tag{6.6-5}$$

Eq. (6.6-4) shows that for $y > 0$ the crystal is stretched in the direction of the x-axis and for $y < 0$ it is compressed. Furthermore, the displacement by the Volterra cut can be seen in Eq. (6.6-5) from the difference in u between $\theta = 0$ and $\theta = 2\pi$.

The Burgers circuit can be expressed formally by a closed line integral. The Burgers vector thus becomes

$$\boldsymbol{b} = \oint \frac{\delta \boldsymbol{u}}{\delta \boldsymbol{s}} \, ds \tag{6.6-6}$$

with the sence of path of the integral being linked to the line sense of the dislocation, forming a right-handed screw.

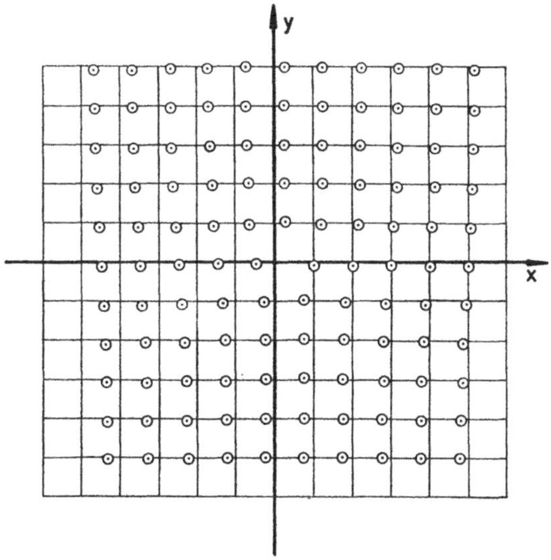

Fig. 6.6/1. Strain field of an edge dislocation calculated from Eqs. (6.6-1) and (6.6-4) ($\nu = 1/3$)

For the calculation of the integral in the above mentioned examples of strain fields, the representation in cylindrical coordinates is chosen with θ variable and r and z constant. $\boldsymbol{b}(b_1\, b_2\, b_3)$ is obtained for the screw dislocation as:

$$\begin{aligned} &b_1 = b_2 = 0 \\ &b_3 = \frac{b}{2\pi} \int_0^{2\pi} d\theta = b \end{aligned} \tag{6.6-7}$$

and for the edge dislocation as:

$$\boldsymbol{b} = (b, 0, 0)\,. \tag{6.6-8}$$

6.7. The Stress Field of a Dislocation

A stress field σ is necessarily coupled with the strain field of the dislocation $\boldsymbol{u}$. Whereas the strain field is a vector field, the stress field is, as indicated in Section 6.2, a tensor field (symmetrical tensor). We consider the same dislocations as in Section 6.6. Since the stress tensor can be described by the force components on the surfaces of a cube, the orientation of this cube has to be fixed. In the case of Cartesian coordinates we choose its edges parallel to the three main axes (Fig. 6.7/1 a); for the representation in cylindrical coordinates, however, the edges are set parallel to the axes r, θ and z (Fig. 6.7/1 b). It might be useful to describe the components by cylindrical coordinates, even in the Cartesian orientation of the cube.

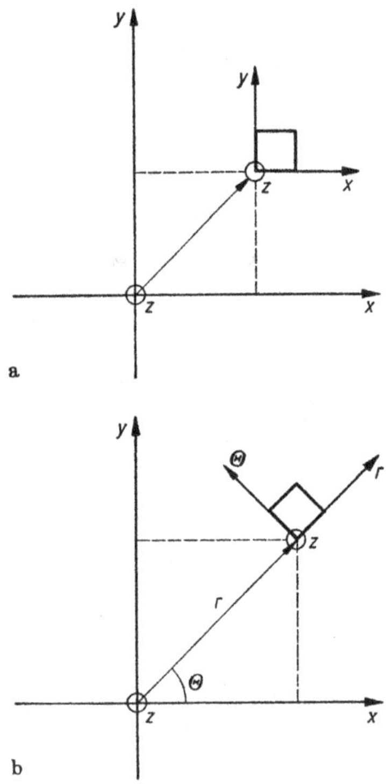

Fig. 6.7/1 a and b. Coordinates of the strain field. a) Orthogonal coordinates, b) cylindrical coordinates

a) Stress Field of the Right-Handed Screw Dislocation

in Cartesian coordinates (cube edges parallel to x, y, z):

The stress tensor is here:

$$\begin{pmatrix} 0 & 0 & \sigma_{13} \\ 0 & 0 & \sigma_{23} \\ \sigma_{31} & \sigma_{32} & 0 \end{pmatrix} \tag{6.7-1}$$

with

$$\begin{aligned}\sigma_{13}=\sigma_{xz}&=-\frac{Gb}{2\pi}\frac{y}{x^2+y^2}\\ \sigma_{23}=\sigma_{yz}&=\frac{Gb}{2\pi}\frac{x}{x^2+y^2}\end{aligned} \tag{6.7-2}$$

G = shear modulus (Section 6.4).

If the cube orientation is given in the direction x, y, z, while the components are indicated in cylindrical coordinates the stress components become:

$$\begin{aligned}\sigma_{13}=\sigma_{xz}&=-\frac{Gb}{2\pi r}\sin\theta\\ \sigma_{23}=\sigma_{yz}&=\frac{Gb}{2\pi r}\cos\theta\,.\end{aligned} \tag{6.7-3}$$

b) Stress Field of the Right-Handed Screw dislocation
in cylindrical coordinates (cube edges parallel to r, θ, z):

Stress tensor:

$$\begin{pmatrix}0 & 0 & 0\\ 0 & 0 & \sigma_{23}\\ 0 & \sigma_{32} & 0\end{pmatrix} \tag{6.7-4}$$

with

$$\sigma_{23}=\sigma_{\theta z}=\frac{Gb}{2\pi r}\,. \tag{6.7-5}$$

A shear stress thus acts on the plane normal to the θ-direction (i.e. on the plane crossing the dislocation line) in the z-direction (or on the z-plane in the θ-direction, respectively) (Fig. 6.7/2). It is proportional to

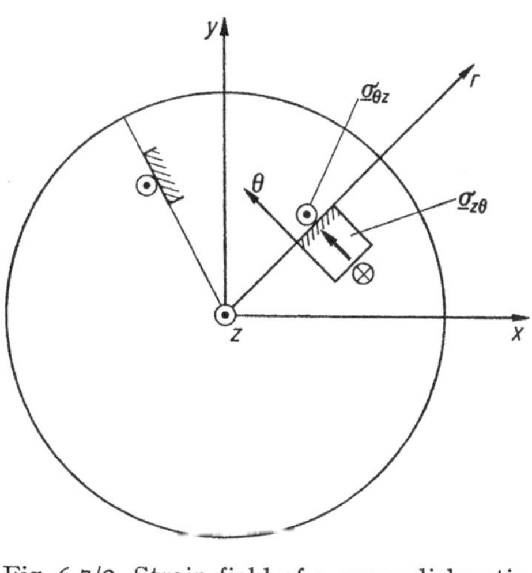

Fig. 6.7/2. Strain field of a screw dislocation

the shear modulus G, and to the Burgers vector b and decreases proportional to $1/r$ with increasing distance from the dislocation.

c) Stress Field of the Edge Dislocation (according to Section 6.6)
in Cartesian coordinates (cube edges parallel to x, y, z):

Stress tensor:

$$\begin{pmatrix} \sigma_{11} & \sigma_{12} & 0 \\ \sigma_{21} & \sigma_{22} & 0 \\ 0 & 0 & \sigma_{33} \end{pmatrix} \tag{6.7-6}$$

$$\begin{aligned} \sigma_{11} &= \sigma_{xx} = -\tau_0\, b \frac{y(3x^2+y^2)}{(x^2+y^2)^2} \\ \sigma_{22} &= \sigma_{yy} = \tau_0\, b \frac{y(x^2-y^2)}{(x^2+y^2)^2} \\ \sigma_{12} &= \sigma_{xy} = \tau_0\, b \frac{x(x^2-y^2)}{(x^2+y^2)^2} \\ \sigma_{33} &= \sigma_{zz} = \nu(\sigma_{xx}+\sigma_{yy}) \\ &= -\tau_0\, b\nu \frac{y(2x^2+y^2)}{(x^2+y^2)^2} \end{aligned} \tag{6.7-7}$$

$\tau_0 = G/2\pi(1-\nu)$,
G = shear modulus,
ν = Poisson's ratio $\approx 1/3$.

Cube parallel to x, y, z, but components in cylindrical coordinates:

$$\begin{aligned} \sigma_{11} &= \sigma_{xx} = -\frac{\tau_0\, b}{r} \sin\theta\,(2+\cos 2\theta) \\ \sigma_{22} &= \sigma_{yy} = \frac{\tau_0\, b}{r} \sin\theta \cos 2\theta \\ \sigma_{12} &= \sigma_{xy} = \frac{\tau_0\, b}{r} \cos\theta \cos 2\theta \\ \sigma_{33} &= \sigma_{zz} = -\frac{\tau_0\, b}{r} 2\nu \sin\theta\,. \end{aligned} \tag{6.7-8}$$

d) Stress Field of the Edge Dislocation
in cylindrical coordinates (cube edges parallel to r, θ, z):

$$\begin{aligned} \sigma_{11} = \sigma_{22} = \sigma_{rr} = \sigma_{\theta\theta} &= -\frac{\tau_0\, b}{r} \sin\theta \\ \sigma_{12} = \sigma_{r\theta} &= +\frac{\tau_0\, b}{r} \cos\theta \\ \sigma_{33} = \sigma_{zz} &= -\frac{\tau_0\, b}{r} 2\nu \sin\theta\,. \end{aligned} \tag{6.7-9}$$

The stress distribution of the edge dislocation in Cartesian coordinates is given in Fig. 6.7/3.

The stress distribution of an arbitrary dislocation with the line lying in the z-axis, results from the superposition of the screw and the edge components. The screw component is

$$b_{\text{screw}} = b \cdot \cos\alpha \tag{6.7-10}$$

and the edge component

$$b_{\text{edge}} = b \cdot \sin\alpha \tag{6.7-11}$$

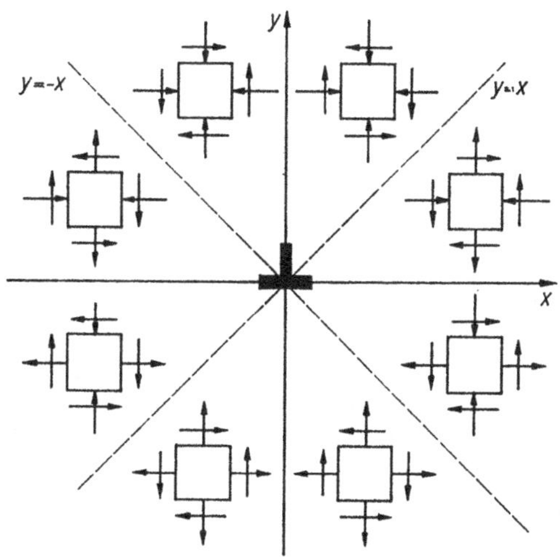

Fig. 6.7/3. Strain field of an edge dislocation

with α being the angle between the Burgers vector and the dislocation line.

The derivation of the stress- and strain field is given in Hirth and Lothe (1968), p. 71.

Chapters 5–6. Summary and Discussion

In Chapters 5 and 6 the concept of the single dislocation, together with the basic definitions were introduced. First we looked at the "skeleton" of the dislocation (line, Burgers vector in 5.2, screw- and edge orientation in 5.3). Next, the dislocation was affected by its surroundings (motion, 5.4; forces on the dislocation, 6.3), and finally it became active (energy, 6.4; line tension, 6.5) and, in its turn, affected the surroundings (strain- and stress fields, 6.6 and 6.7).

The following points should be emphasized:

a) Isotropic Elastic Medium. Up till now the crystalline character of matter has hardly been stressed. Only in the definition of the Burgers vector was it mentioned that the Burgers vector has to be a translation vector of the crystal lattice. All the other considerations are also valid for isotropic elastic materials which contain a dislocation introduced by a Volterra cut. Therefore, the basis is pure isotropic elasticity theory. Some modifications are, of course, possible and occasionally even necessary, and these can be made by the application of the elasticity theory of anisotropic media.

b) Pseudo-Vector Character of the Burgers Vector. The Burgers vector is invariant along a dislocation in magnitude, orientation, and sign, independent of the curves in the dislocation line, as long as the line sense is uniformly attributed to the line. If the line sense is inverted, the sign of the Burgers vector is also inverted. Its magnitude and orientation, however, remain unchanged. The sign of the Burgers vector can therefore only be given together with the line sense. The coupling between the sign of the Burgers vector and the line sense has to be taken into account in the correct formulation of geometrical statements, such as, for example, the movement rule [Eq. (5.4-1)] and the Peach and Koehler formula [Eq. (6.3-4)]. Only then is a consistent formulation possible.

This invariance is guaranteed only if the Burgers vector is considered as an "abstract" pseudo-vector, i.e. as a coordinate difference, as demonstrated by Frank by transcription on to the perfect reference lattice. Only then is the Burgers vector independent of the local crystal orientation and strain. Hence, the expression "local Burgers vector" should be avoided. It might be replaced by "local lattice vector".

c) The force always acts normal to the dislocation line. At first it does not seem obvious that, in spite of a directed shear stress $\boldsymbol{\tau}$ and a directed displacement $\boldsymbol{b}$, the dislocation loop under stress should expand uniformly in all directions, even against $\boldsymbol{\tau}$ and $\boldsymbol{b}$. We have to remember that a dislocation line is basically a *boundary line* between a "displaced" and an "undisplaced" part of the crystal [see movement rule (5.4-1)]. The motion of a boundary line only makes sense in a direction vertical to its orientation. At the border between two countries, for example, nothing would be changed if all the border stones were moved within the direction of the boundary line; besides, in our case the border stones, i.e. the atoms, remain practically in their positions while a dislocation moves. A force component (work per unit distance) in the direction of the line is therefore meaningless. The Peach and Koehler formula automatically eliminates the tangential component of the force.

d) The energy per unit length of the dislocation is partly concentrated in the dislocation core; the greater part of it, however, is spread over the surroundings, i.e. the stress field of the dislocation, and extends, in principle, over the whole crystal. It diverges for a crystal as r tends to infinity. An exact delimitation of the energies of individual dislocations is therefore not possible. The arbitrary definition of the effective radius allows estimations to be made.

For the dislocation theory in general a kind of "uncertainty principle" seems to apply. One either simplifies the formulae employed, thus leading to inaccuracy, or one tries to carry out the calculation as exactly as possible. In the latter case, however, the calculations involve a number of parameters which are either only approximately known, or which have to be estimated, and therefore inaccurate results are also obtained.

Suggestions for Problems

Chapters 5 and 6 have been kept short in this book since most of the problems described are given in greater detail in other books, e.g. Read (1953), Weertman and Weertman (1964) and Hirth and Lothe (1968). These books also suggest many problems.

It might be particularly useful to practise the use of the movement rule and the Peach and Koehler formula, and to become familiar with the pseudo-vector character of the Burgers vector.

7. The Interaction of Dislocations

7.1. Outline

Up to the present stage the properties of an individual dislocation in a crystal have been investigated. The knowledge of the stress field of the dislocation makes it possible now to understand the interactions between two dislocations. This leads to the study of dislocation reactions, i.e. the nodes formed by dislocations. At the end of the chapter dislocation reactions in special crystal structures will be investigated.

7.2. Interaction between Two Parallel Screw Dislocations

The Peach and Koehler formula (6.3-4) indicates the force which acts on a dislocation in a stress field. This stress field can originate either from external forces or from other dislocations. We consider the interaction between two straight parallel dislocations of the same type. One of them shall lie in the z-axis.

In the first example we deal with two right-handed screw dislocations. In analogy to the considerations of the forces between charged particles in electrostatics we assume the dislocation in the z-axis to be the source of the stress field and investigate the action of the force due to this stress field on the other dislocation. The Burgers vector and line sense may point in the direction of the positive z-axis. We can here directly apply the representation in polar coordinates. $\boldsymbol{\sigma}_b$ thus becomes [Eq. (6.7-5)]:

$$\boldsymbol{\sigma}_b = \boldsymbol{\sigma}_{z\theta} = \frac{Gb}{2\pi r} \boldsymbol{u}_\theta . \tag{7.2-1}$$

Together with Eq. (6.3-4) the force per unit length of the dislocation becomes:

$$\boldsymbol{f} = |\boldsymbol{b}| \cdot \frac{Gb}{2\pi r} [\boldsymbol{u}_\theta \times \boldsymbol{u}_z] = \frac{Gb^2}{2\pi r} \boldsymbol{u}_r \tag{7.2-2}$$

$\boldsymbol{u}_r$, $\boldsymbol{u}_\theta$, $\boldsymbol{u}_z$ = unit vectors in the direction of the respective coordinate axes.

Hence, the force between two equal screw dislocations is repelling and decreases $\propto 1/r$.

If the dislocation at the position $\boldsymbol{r}$ is of the left-handed screw type, $\boldsymbol{u}_z$ and $\boldsymbol{f}$ change their signs and the two dislocations of opposite signs now attract each other.

If the first dislocation, lying in the position of the z-axis, has right-handed screw character with a Burgers vector $\boldsymbol{b}_1$ and the other *parallel* dislocation with $\boldsymbol{b}_2$ is of mixed character, the force between them is (Hirth and Lothe, 1968, p. 117):

$$\boldsymbol{f} = \frac{G(\boldsymbol{b}_1 \cdot \boldsymbol{l}_1)(\boldsymbol{b}_2 \cdot \boldsymbol{l}_2)}{2\pi r} \boldsymbol{u}_r. \tag{7.2.-3}$$

$\boldsymbol{l}_1$ and $\boldsymbol{l}_2$ being the unit vectors in the line sense. Eq. (7.2-3) fulfills the invariance condition (5.2-1): $F(\boldsymbol{l}, \boldsymbol{b}) = F(-\boldsymbol{l}, -\boldsymbol{b})$.

7.3. Interaction between Two Parallel Edge Dislocations

Again we assume one edge dislocation lying in the z-axis acting as the source of the stress field. The other edge dislocation may be in an arbitrary position, but parallel to the first. Both dislocations are located such that the additional half plane [Eq. (5.3-1)] points to the side of the positive y-axis. The Burgers vectors of the two dislocations are assumed to be equal. First we consider the case of the second dislocation being able to glide on a plane parallel to the xy-plane, but not to climb. Hence, the force on the second dislocation has to point in the x-direction and, according to the Peach and Koehler formula [Eq. (6.3-4)] it is produced by the shear stress component σ_{xy}.

$$f_x = b \cdot \sigma_{xy} = \tau_0 \, b^2 \frac{x(x^2 - y^2)}{(x^2 + y^2)^2} \tag{7.3-1}$$

with

$$\tau_0 = \frac{G}{2\pi(1-\nu)}. \tag{7.3-2}$$

The following segments can be distinguished (Fig. 7.3/1):

1. $x > 0$ $\quad |x| > |y| \quad f_x > 0$
2. $x > 0$ $\quad |x| < |y| \quad f_x < 0$
3. $x < 0$ $\quad |x| < |y| \quad f_x > 0$
4. $x < 0$ $\quad |x| > |y| \quad f_x < 0$.

This shows that, in the segments 1 and 4 the force is such that the dislocations repel one another, whereas they attract one another is segments 2 and 3. Thus, for two equal parallel edge dislocations there is a stable situation for glide at $x = 0$. In the case of two edge dislocations with opposite signs stable situations also exist, but here they appear for $|x| = |y|$.

The physical reason for the stability of the arrangement of two dislocations placed on top of each other lies in the fact that they form part of a possible interface between two crystals, slightly tilted with respect to one another (Section 11.3).

If, in addition to glide, any other motion of the dislocations is allowed, dislocations of the same kind repel one another whereas dislocations of opposite kind attract one another. The force vectors, however, do not necessarily lie in the plane defined by the two dislocations. If the dislocations are in an arbitrary orientation with respect to one another, the effect of every element of one dislocation on every element of the other has to be calculated (Hirth and Lothe, 1968, p. 113).

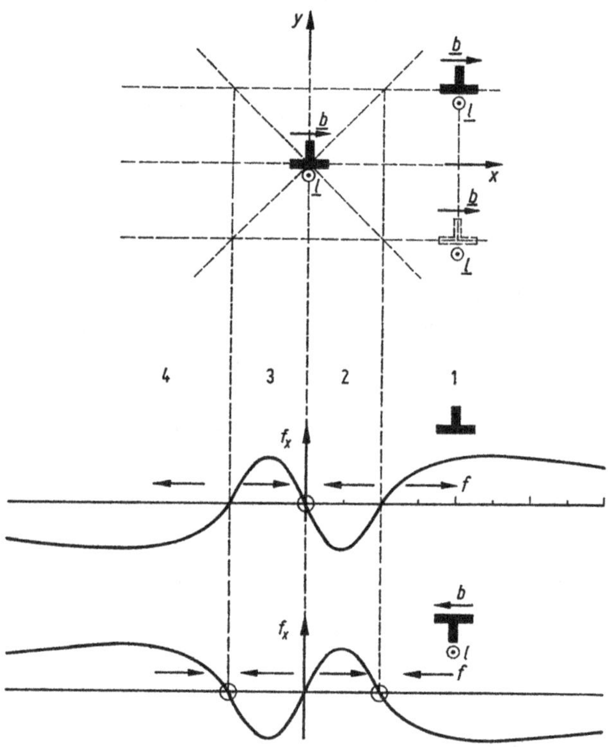

Fig. 7.3/1. Force between two parallel edge dislocations [Eq. (7.3-1)] under the condition that only glide is permitted. (The circles mark stable positions)

7.4. Dislocation Reactions

If two dislocations attract each other, a *dislocation reaction* may take place with the formation of a single dislocation. It does not necessarily happen over the whole length of the dislocations, so that dislocations can bifurcate and form nodes. In these cases, the following condition stated by F. C. Frank applies. If we consider the dislocation segments with their appropriate line sense and the Burgers vectors with the corresponding signs, Frank's node condition states:

The sum of the Burgers vectors flowing into the node (line sense!) equals the sum of the Burgers vectors flowing out of the node.

$$\boxed{\sum_i \boldsymbol{b}_i \,(\text{in}) = \sum_j \boldsymbol{b}_j \,(\text{out})}\,. \tag{7.4-1}$$

This node condition is a law of conservation of Burgers vectors (continuity), analogous to the conservation of electrical currents in a node of conductors.

The node condition can also be formulated as follows:

If the line senses, and accordingly the signs of Burgers vectors are chosen such that all the dislocations flow into the node, then the sum of the Burgers vectors is zero.

The node condition can be proved as shown in Fig. 7.4/1, by moving a Burgers circuit along the dislocations; the total Burgers vector has to remain unchanged as long as no dislocation is cut.

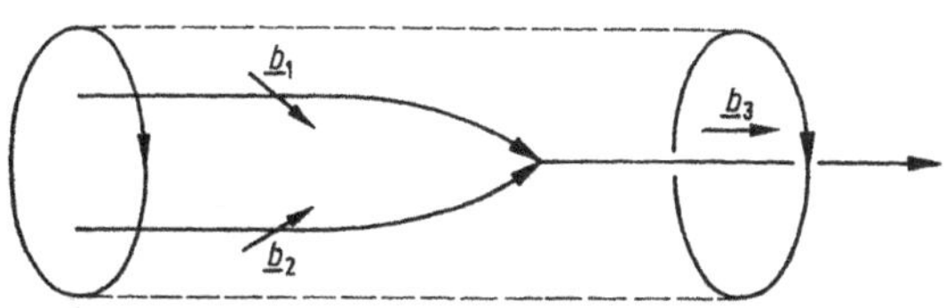

Fig. 7.4/1. The proof of Frank's node condition. (Displacement of the Burgers loop)

If a dislocation reaction occurs, the node condition furnishes the balance of the Burgers vectors; it does not indicate, however, *whether* this reaction will take place or not. This is determined by the increase or decrease in energy of the crystal during the reaction. Neglecting all the other influences, the energy per unit length of a dislocation is approximately proportional to b^2 [Eq. (6.4-10)]. According to this criterion two dislocations 1 and 2 will combine to give one dislocation 3, if

$$b_1^2 + b_2^2 > b_3^2. \tag{7.4-2}$$

From the node condition, it follows that

$$\boldsymbol{b}_1 + \boldsymbol{b}_2 = \boldsymbol{b}_3. \tag{7.4-3}$$

This means, geometrically, that the vectors $\boldsymbol{b}_1$, $\boldsymbol{b}_2$ and $\boldsymbol{b}_3$ together have to form an acute-angled triangle (Fig. 7.4/2a). An equilibrium exists if

$$b_1^2 + b_2^2 = b_3^2 \tag{7.4-4}$$

i.e. if the three Burgers vectors form a right-angled triangle (Fig. 7.4/2b). In the case of

$$b_1^2 + b_2^2 < b_3^2 \tag{7.4-5}$$

the Burgers vectors form an obtuse-angled triangle. A dislocation reaction would result here in an increase in the energy of the crystal, so it is not favoured (Fig. 7.4/2c).

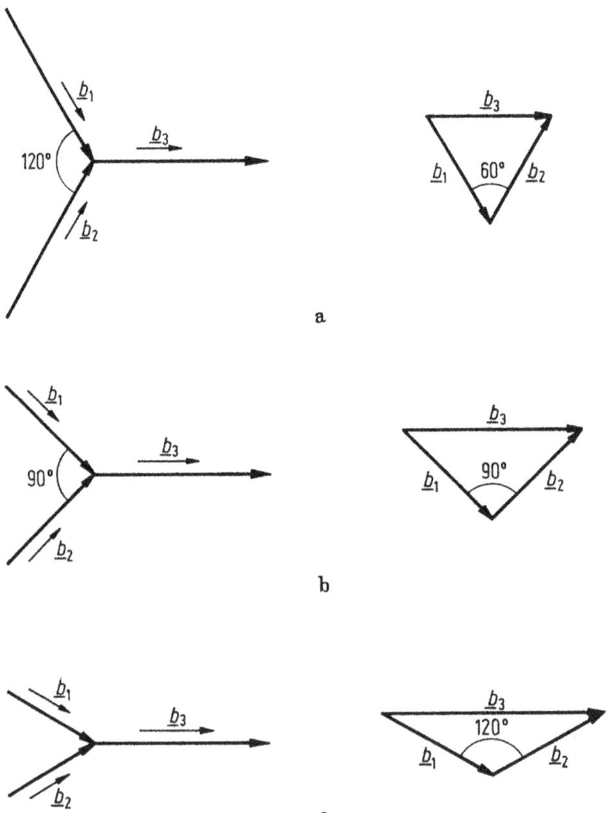

Fig. 7.4/2a–c. The relative stability of dislocation reactions: a) stable, b) labile, c) unstable

The energy difference between a screw dislocation and an edge dislocation has not been considered in these estimations. Taking this difference into account [Eq. (6.4-12)], the limiting angle for the stability in the Burgers vector triangle may differ from 90°.

7.5. Dislocation Reactions in Special Crystal Structures

From the geometrical definition of the Burgers vector it follows that it must be a translation vector of the crystal. Moreover, the energy considerations [$\Gamma \sim b^2$, Eq. (6.4-10)] show that the most favorable Burgers vector is the smallest possible translation vector, since the energy of a dislocation with a double Burgers vector would be four times that of the simple dislocation. The smallest possible translation vectors in different crystal structures are given in Sections 3.3 etc.

A discussion of several typical examples follows:

Face-Centred Cubic (FCC) Structure

1. The basic reaction is here of the type (Fig. 7.5/1):

$$(1/2)\,[\bar{1}\,1\,0] + (1/2)\,[0\,\bar{1}\,1] = (1/2)\,[\bar{1}\,0\,1]^{*} \tag{7.5-1}$$

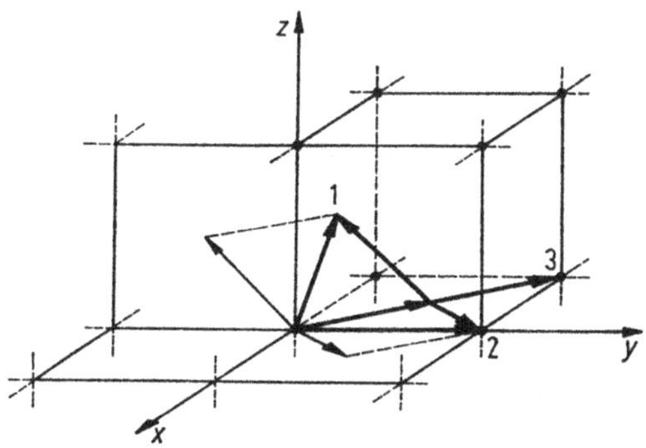

Fig. 7.5/1. Dislocation reactions in the f.c.c. structure. *1.* Stable, *2.* labile, *3.* unstable

with the b^2-balance ($= \boldsymbol{b}^\top \boldsymbol{G} \boldsymbol{b}$):

$$a^2/2 + a^2/2 > a^2/2. \tag{7.5-2}$$

The reaction will thus happen in the "direction of the arrow".

2. A situation where the reaction may go either way is:

$$(1/2)\,[\bar{1}\,1\,0] + (1/2)\,[1\,1\,0] = [0\,1\,0] \tag{7.5-3}$$

with the b^2-balance

$$a^2/2 + a^2/2 = a^2. \tag{7.5-4}$$

* We write the Burgers vectors in dimensionless coordinates, i.e. $(1/2)\,[\bar{1}\,1\,0]$ instead of $(a/2)\,[\bar{1}\,1\,0]$. The lattice constant, a will appear for $|b|^2$ through the metric tensor $\boldsymbol{G}$. This notation is of special advantage for non-orthogonal systems.

3. The following reaction will only occur with difficulty, since it creates an unstable situation:

$$(1/2)\ [\bar{1}\,1\,0] + (1/2)\ [\bar{1}\,1\,0] = [\bar{1}\,1\,0] \tag{7.5-5}$$

with the b^2-balance:

$$a^2/2 + a^2/2 < 2a^2. \tag{7.5-6}$$

The corresponding reactions also apply for the rock salt and the diamond structure, as these two structures contain the same translation unit as the face-centred cubic structure.

Hexagonal Structure

The hexagonal structure has a preferential plane, the basic plane (001). The smallest translation vectors are of the type [100]. A possible reaction is as follows (Fig. 7.5/2):

$$[1\,0\,0] + [0\,1\,0] = [1\,1\,0] \tag{7.5-7}$$

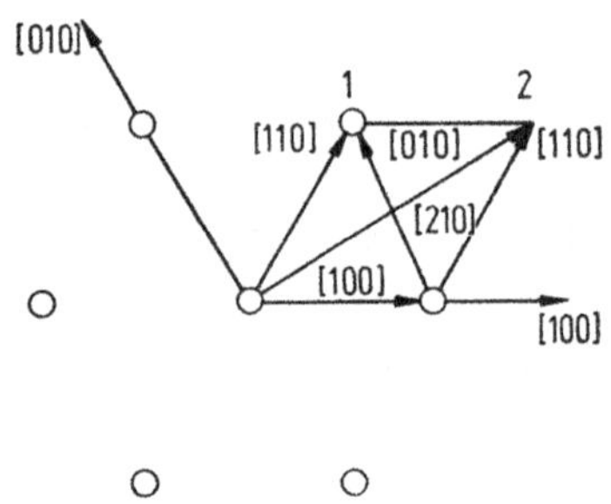

Fig. 7.5/2. Dislocation reactions in the c.p.h. structure. *1.* Stable, *2.* unstable

with the b^2-balance [with G from Eq. (3.7-1)]:

$$a^2 + a^2 > a^2 \tag{7.5-8}$$

and an improbable reaction:

$$[1\,0\,0] + [1\,1\,0] = [2\,1\,0] \tag{7.5-9}$$

$$a^2 + a^2 < 3\,a^2. \tag{7.5-10}$$

In several hexagonal materials glide is, in principle, also possible on "prism planes" (100) and "pyramid planes" (102); however, the Burgers vectors in the c-direction become very large.

The corresponding reactions in the basal plane occur also in graphite. However, because of the pronounced layer structure of graphite, the dislocation reactions in prism- and pyramid planes are practically excluded.

Body-Centred Cubic Structure

In the bcc structure the smallest Burgers vectors are of the type (1/2) ⟨1 1 1⟩, the next larger ones of type ⟨1 0 0⟩ and even larger ones of type ⟨1 1 0⟩.

The most important dislocation reaction is of the following type:

$$(1/2)\ [1\,1\,1] + (1/2)\ [\bar{1}\,\bar{1}\,1] = [0\,0\,1] \tag{7.5-11}$$

with the b^2-balance:

$$(3/4)\ a^2 + (3/4)\ a^2 > a^2. \tag{7.5-12}$$

The reaction

$$(1/2)\ [1\,1\,1] + (1/2)\ [1\,1\,\bar{1}] = [1\,1\,0] \tag{7.5-13}$$

with

$$(3/4)\ a^2 + (3/4)\ a^2 < 2a^2 \tag{7.5-14}$$

is unstable (Fig. 7.5/3). In certain rare cases small segments of ⟨1 1 0⟩-dislocations have been observed (Hale and Brown, 1968).

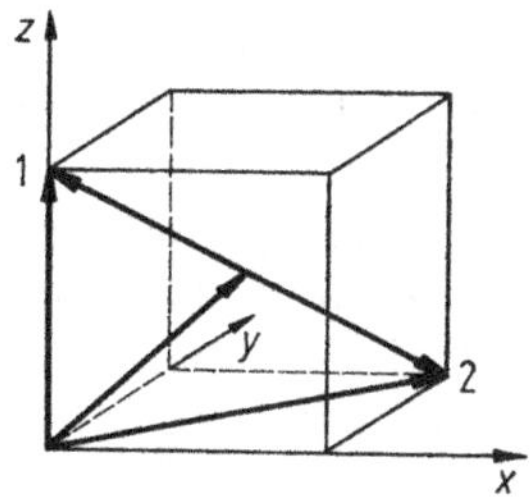

Fig. 7.5/3. Dislocation reactions in the b.c.c. structure. *1.* Stable, *2.* unstable

8. Partial Dislocations

8.1. Outline

We have seen that the Burgers vectors are the smallest possible translation vectors of the lattice. When a dislocation moves through the crystal, the Burgers vector represents a displacement (of one part of the crystal with respect to the other) such that the crystal which was perfect before the passage of the dislocation is again perfect afterwards.

In close-packed and related structures this shift may become divided into two steps, and in special layer structures even into more than two. The passage of such "partial steps" or "partial dislocations" (abbreviated to "partials") involves the formation of stacking faults. Partial dislocations thus appear as boundaries of stacking faults. In the following we consider the situation in the face-centred cubic and in the hexagonal structures.

In the fcc structure the stacking sequence of partial dislocations is not arbitrary. The determination of this sequence will be discussed in Section 8.3. As an example of the application of the considerations on partials, the phase transformation of cobalt will be discussed in Section 8.4.

8.2. Partial Dislocations in the Face-Centred Cubic Structure

We consider the face-centred cubic structure as the ABC-stacking of close-packed layers of spheres (Section 3.3) and assume that a dislocation glides between a B- and a C-layer such that the lower layers (A and B) remain fixed and the C-layer is shifted (vector $\boldsymbol{m}$ of the movement rule (5.4-1) points upwards) (Fig. 3.2/3). As the dislocation sweeps over the B-plane, a C-atom will be shifted from one C-position into the next. This shift can take place in two steps, the first step leading from a C-position into an A-position, and the second from this A-position into the next C-position. Steps of this kind are, for example (Fig. 8.2/1 — solid arrows):

$$\begin{array}{ccccc} (1/6)\,[\bar{2}\,1\,1] & + & (1/6)\,[\bar{1}\,2\,\bar{1}] & = & (1/2)\,[\bar{1}\,1\,0] \\ (1) & & (2) & & (3) \\ C_0 \rightarrow A & + & A \rightarrow C_1 & = & C_0 \rightarrow C_1. \end{array} \qquad (8.2\text{-}1)$$

The left hand side of the equation represents the Burgers vectors of the two partials, which together form the total dislocation with the Burgers vector (1/2) $[\bar{1}\,1\,0]$.

The b^2-balance of this reaction is:

$$a^2/6 + a^2/6 < a^2/2. \tag{8.2-2}$$

From this point of view the splitting of partial dislocations is favoured, i.e. they repel each other. It can also be seen from Fig. 8.2/1 that the corresponding Burgers vectors form an obtuse-angled triangle.

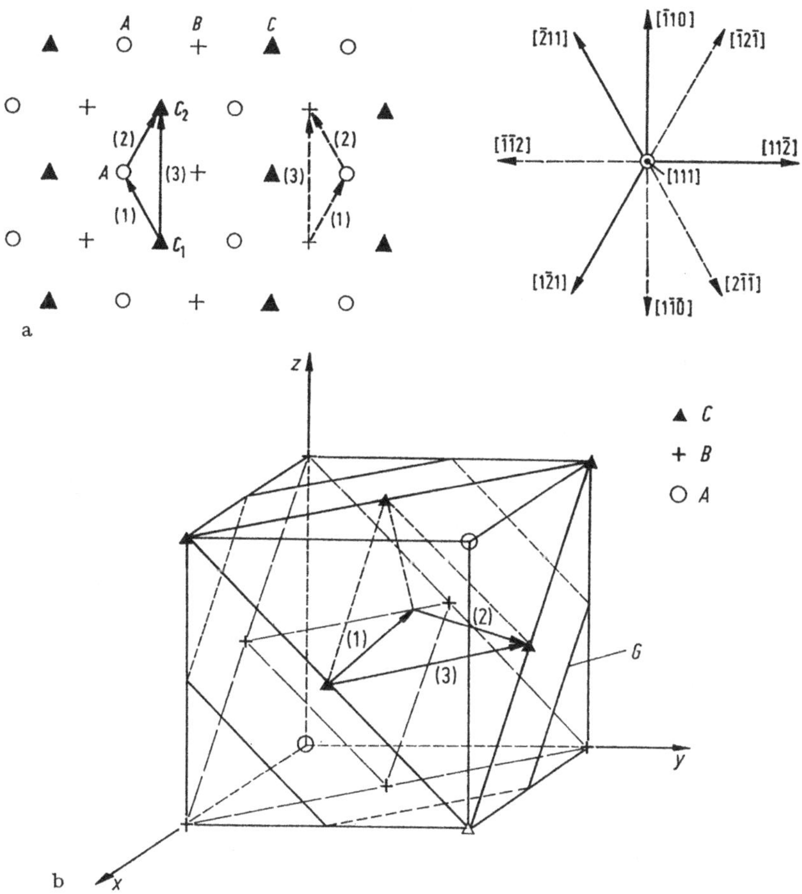

Fig. 8.2/1 a and b. Partial dislocations in the f.c.c. structure. Full arrows: displacement of the part above the plane of the paper; dashed arrows: displacement below the plane of the paper. a) projection in the [1 1 1]-direction. (The arrows in the right-hand diagram indicate the crystallographic directions.) b) situation in space (*G* glide plane)

Between the two partial dislocations, a stacking fault is formed which, in its turn, increases the energy of the crystal. The two partial dislocations are held together by the stacking fault ribbon.

The splitting of a perfect dislocation into two partials can be estimated as follows: The force per unit length between two screw dislocations [Eq. (7.2-3)] is balanced by the change in total energy of the stacking fault, i.e. by the change in the surface of the stacking fault [times the stacking fault energy γ (Section 3.10)] on changing the spacing r of the two partials.

$$f \cdot L \approx L \frac{G(\boldsymbol{b}_1 \cdot \boldsymbol{l}_1)(\boldsymbol{b}_2 \cdot \boldsymbol{l}_2)}{4\pi\, \alpha\, r} \approx \frac{d(L \cdot \gamma \cdot r)}{dr} = L\gamma \qquad (8.2\text{-}3)$$

$$r \approx \frac{G(\boldsymbol{b}_1 \cdot \boldsymbol{l}_1)(\boldsymbol{b}_2 \cdot \boldsymbol{l}_2)}{4\pi\, \alpha\, \gamma} \qquad (8.2\text{-}4)$$

with

$1 < \alpha < (1-\nu)$ = correction for the deviation from the screw orientation. The more accurate formula

$$r = \frac{G}{2\pi\gamma}\left[(\boldsymbol{b}_1 \cdot \boldsymbol{l}_1)(\boldsymbol{b}_2 \cdot \boldsymbol{l}_2) + \frac{[\boldsymbol{b}_1 \times \boldsymbol{l}_1] \cdot [\boldsymbol{b}_2 \times \boldsymbol{l}_2]}{1-\nu}\right] \qquad (8.2\text{-}5)$$

is derived in Hirthe and Lothe (1968) p. 298. Fig. 8.2/2 shows a split dislocation. (For the numerical values of the stacking fault energies see remark in Section 3.10.)

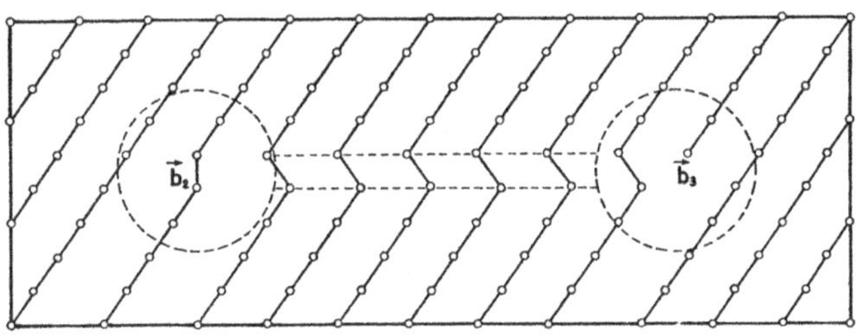

Fig. 8.2/2. Stacking sequence of a dislocation split into partials

Another possibility for the splitting of a perfect dislocation is:

$$(1/2)\,[1\,0\,1] = (1/6)\,[1\,\bar{2}\,1] + (1/3)\,[1\,1\,1].$$

According to the b^2-criterion

$$a^2/2 = a^2/6 + a^2/3$$

this reaction is doubtful. The Burgers vector (1/3) [1 1 1] does not lie in a ⟨1 1 1⟩-glide plane. Such a dislocation can therefore not glide. It is called a "sessile" partial in contrast to a "glissile" partial. The sessile partials are also called Frank partials and the glissile $(1/6)\langle 1\,1\,2\rangle$-partials are termed Shockley partials. Frank partials occur where layers are inserted or removed (condensation of interstitial atoms or vacancies).

Several possible stacking fault configurations were discussed in Section 3.9. The boundary lines of such configurations inside the crystal are various types of partial dislocations.

8.3. The Sequence of Partial Dislocations

Eq. (8.2-1) shows that the displacement caused by a total dislocation can be split up into two partial steps. After the first partial has passed, the stacking order has changed from

$$\begin{array}{c} \downarrow \\ ABCABCABC \\ KKKKKKK \end{array} \tag{8.3-1}$$

to

$$\begin{array}{c} \downarrow \\ ABCAB\boldsymbol{ABCA} \\ KKK\boldsymbol{HH}KK. \end{array} \tag{8.3-2}$$

The boldface part is that region of the crystal that has moved. The inversion of the sequence of the partials would, after the first step, produce the following stacking order:

$$\begin{array}{c} \downarrow \\ ABCAB\boldsymbol{BCABC} \\ KKK\boldsymbol{LL}KKK \end{array} \tag{8.3-3}$$

i.e. the labile sequence L ($=ABB$, etc.) would occur, with two close-packed layers of spheres on top of one another. This shows that the sequence of partial dislocations is not arbitrary, i.e. the vector addition of the Burgers vectors of the partials is not commutative. In Fig. 8.3/1

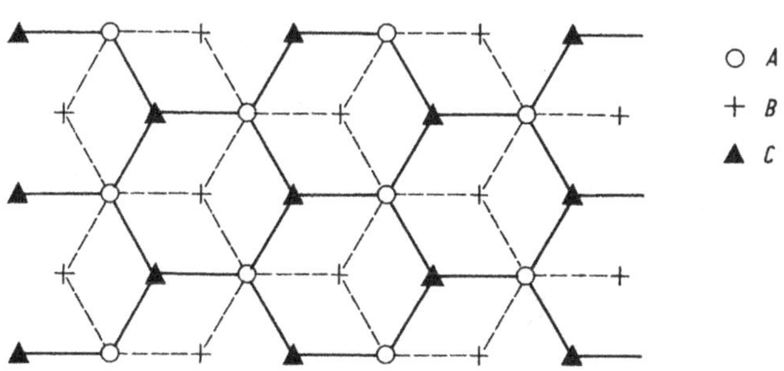

Fig. 8.3/1. Possible displacements by partial dislocations. Full lines: upper part displaced, dashed lines: lower part displaced

all the possible steps on the (1 1 1) glide plane that are allowed for the upper half of the crystal are indicated by solid lines. With respect to the movement rule (5.4-1), these are the steps for which $\boldsymbol{m}$ points in the [1 1 1]-direction. All these steps together form a hexagonal arrangement (analogous to the graphite structure, Section 4.9). If $\boldsymbol{m}$ points in the opposite ($[\bar{1}\bar{1}\bar{1}]$-) direction, the dotted steps (Figs. 8.2/1 and 8.3/1) become possible. In the first case the stacking order changes from (8.3-1) to (8.3-2). However, if the "bottom" part is displaced and the "top" part kept fixed, then the stacking order after the first step is

$$\begin{array}{c}\boldsymbol{CABCA}CABCA\\ KKK\boldsymbol{HH}KKK.\end{array} \tag{8.3-4}$$

The splitting of dislocations therefore depends on the sign of $\boldsymbol{m}$.

Four equivalent {1 1 1} glide planes exist in the cubic structure and we must now determine the stacking sequence of the partial dislocations in all these glide planes. The four {1 1 1} planes can be joined to a tetrahedron, the "Thompson's tetrahedron" (Thompson, 1953). The edges of this tetrahedron are formed by the Burgers vectors of the type 1/2⟨1 1 0⟩, of the perfect dislocations, and the connections from the corners to the face centres of the triangles are composed of the Burgers vectors of the partial dislocations of the type 1/6⟨1 1 2⟩. In order to fulfill the conditions for the splitting of dislocations and also to meet the movement rule, $\boldsymbol{m}$ has to point from the respective tetrahedral surfaces towards the *interior* of the tetrahedron. Otherwise the partial Burgers vectors, taken in the correct order, would externally surround the triangle of the tetrahedral surface. Thus, two possible tetrahedra exist for the eight different ⟨1 1 1⟩ orientations of $\boldsymbol{m}$ (Figs. 8.3/2–4). The total Burgers

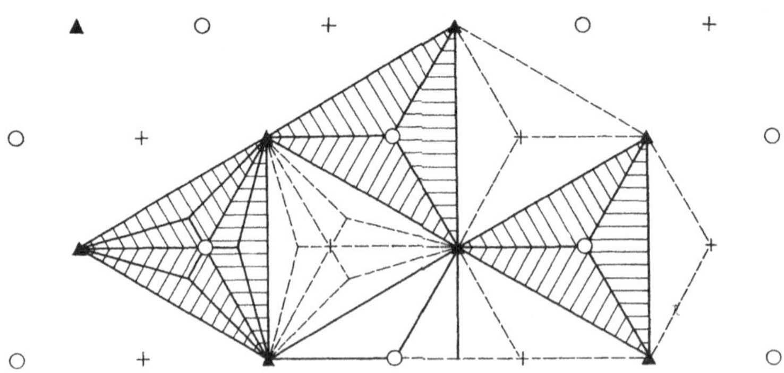

Fig. 8.3/2. Composition of Thompson's tetrahedra for the representation of dislocation splitting on all four {1 1 1}-glide planes. $\boldsymbol{m}$ has to point *into* the tetrahedra

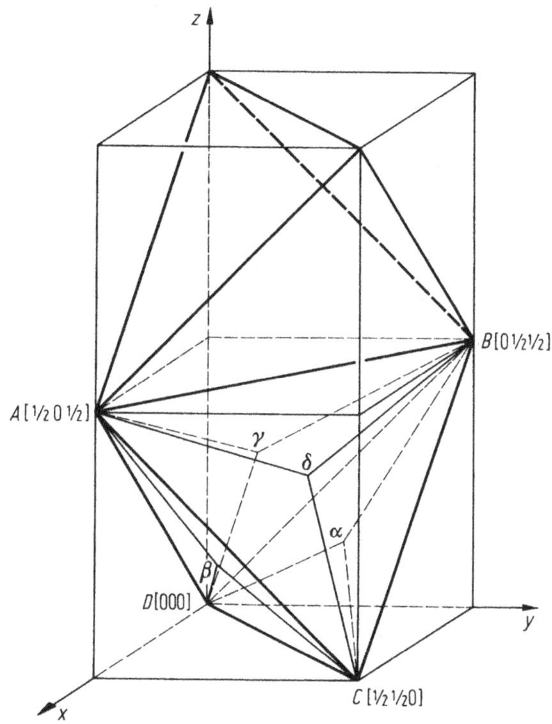

Fig. 8.3/3. Position of Thompson's tetrahedra in the cubic elementary cell

vector $\boldsymbol{b}$ results from the addition of the partial Burgers vectors $\boldsymbol{b}^{(1)}$ and $\boldsymbol{b}^{(2)}$; the first term of the vector addition, $\boldsymbol{b}^{(1)}$ is assigned to that partial which points in the direction of $\boldsymbol{v}$ of the movement rule [Eq. (5.4-1)]. $\boldsymbol{l}$ and $\boldsymbol{v}$ can always be defined in such a way that $\boldsymbol{m}$ corresponds to one of the four $\langle 1\,1\,1\rangle$ directions of Fig. 8.3/4. When, on splitting the "perfect" Burgers vector $\boldsymbol{b}$, the partial $\boldsymbol{b}^{(1)}$ points to the centre of a tetrahedral face, the sequence of partial dislocations can be determined geometrically in an unambiguous manner by means of the Thompson tetrahedron.

Here follows a description of an algebraic procedure for determining the sequence of partials. This procedure can be useful, for example, in connection with computer calculations (Bollmann and Perry, 1967).

We define:

the line sense $\boldsymbol{l}$,

the (total) Burgers vector $\boldsymbol{b}$ [of the type $(1/2)\langle 1\,1\,0\rangle$],

the glide plane given by $\boldsymbol{m}$, of the form $\langle 1\,1\,1\rangle$.

The different steps are:

1. An auxiliary vector $\boldsymbol{c}$ is calculated

$$\boldsymbol{c} = (1/3)\ [\boldsymbol{b} \times \boldsymbol{m}]. \tag{8.3-5}$$

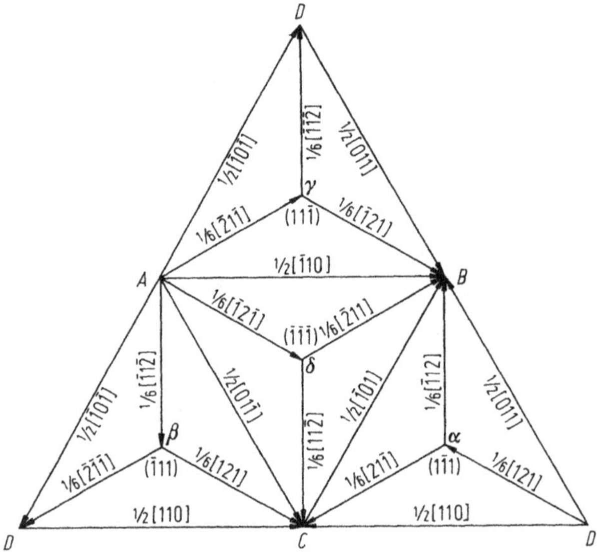

Fig. 8.3/4. Thompson's tetrahedron net

It is of the type $(1/6)\langle 1\,1\,2\rangle$.

2. The vectors $\boldsymbol{x}$ and $\boldsymbol{y}$ are determined:

$$\boldsymbol{x} = (1/2)\,(\boldsymbol{b} + \boldsymbol{c}) \qquad \boldsymbol{y} = (1/2)\,(\boldsymbol{b} - \boldsymbol{c}). \tag{8.3-6}$$

$\boldsymbol{x}$ and $\boldsymbol{y}$ are the Burgers vectors of the two partials, but it is not yet known which is $\boldsymbol{b}^{(1)}$ (attributed to the line $\boldsymbol{l}^{(1)}$ leading in the direction $\boldsymbol{v}$), with

$$\boldsymbol{v} = [\boldsymbol{m} \times \boldsymbol{l}] \tag{8.3-7}$$

[from cyclic permutation of Eq. (5.4-1)].

3. The vector $\boldsymbol{B}$ is defined as:

$$\boldsymbol{B} = [\boldsymbol{b}^{(1)} \times \boldsymbol{b}^{(2)}]. \tag{8.3-8}$$

It can be parallel or antiparallel to $\boldsymbol{m}$, hence the $\operatorname{sgn}(\boldsymbol{B} \cdot \boldsymbol{m})$ depends on the sequence of $\boldsymbol{b}^{(1)}$ and $\boldsymbol{b}^{(2)}$. This sign is determined by:

$$\underline{\operatorname{sgn}(\boldsymbol{B} \cdot \boldsymbol{m}) = \operatorname{sgn}\Big(\prod_i c_i\Big) \cdot \operatorname{sgn}\Big(\prod_i m_i\Big).} \tag{8.3-9}$$

The two factors on the right are the signs of the products of the coordinates of $\boldsymbol{c}$ and $\boldsymbol{m}$. Thus, with $\boldsymbol{m}$ and $\boldsymbol{c}$ given, $\boldsymbol{x}$ and $\boldsymbol{y}$ have to be attributed to $\boldsymbol{b}^{(1)}$ and $\boldsymbol{b}^{(2)}$ such that Eq. (8.3-9) is fulfilled.

As the vectors $\boldsymbol{m}$ and $\boldsymbol{B}$ are parallel and hence, their coordinates are proportional, $\operatorname{sgn}(\boldsymbol{B}\cdot\boldsymbol{m})$ can be checked on a single coordinate, e.g.

$$\begin{aligned}\operatorname{sgn}(\boldsymbol{B}\cdot\boldsymbol{m}) &= \operatorname{sgn}(B_3\, m_3)\\ &= \operatorname{sgn}(b_1^{(1)}\, b_2^{(2)} - b_2^{(1)}\, b_1^{(2)})\cdot\operatorname{sgn}(m_3)\,.\end{aligned} \tag{8.3-10}$$

Example

We assume a dislocation with

$$\boldsymbol{l} = [1\,2\,1]\,, \qquad \boldsymbol{b} = (1/2)\,[1\,1\,0]\,.$$

The screw component of this dislocation is right-handed [$\operatorname{sgn}(\boldsymbol{l}\cdot\boldsymbol{b}) = (+)$, Eq. (5.3-4)] and the extra plane of the edge component ($\boldsymbol{p} = [\boldsymbol{l}\times\boldsymbol{b}]$, Eq. (5.3-3)) is on the side $[\bar{1}\,1\,\bar{1}]$. This direction remains normal to the glide plane and we choose $\boldsymbol{m} = [1\,\bar{1}\,1]$ (Fig. 8.3/5). According to Eq. (8.3-7)

$$\boldsymbol{v} = [\boldsymbol{m}\times\boldsymbol{l}] = [\bar{1}\,0\,1]$$

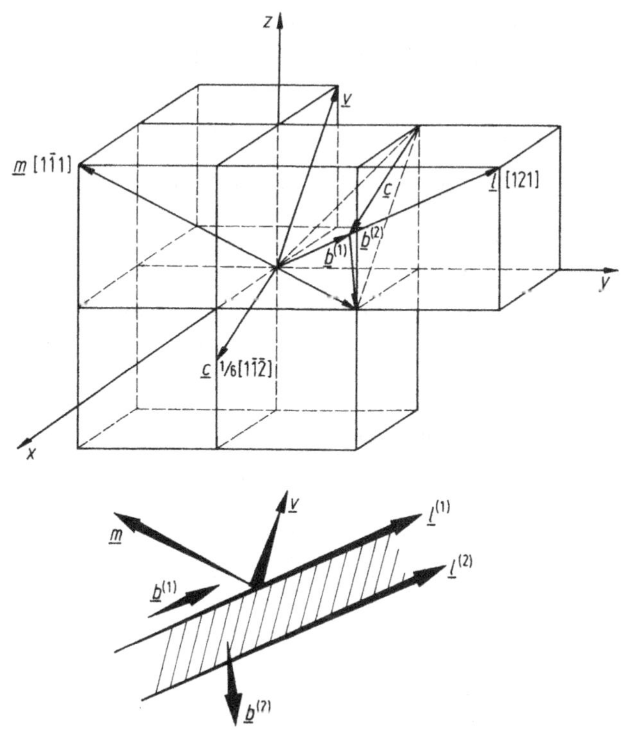

Fig. 8.3/5. Example of the splitting of a dislocation

and $\boldsymbol{l}^{(1)}$ is the partial dislocation which leads in the direction of $\boldsymbol{v}$. From Eq. (8.3-5)

$$\boldsymbol{c} = (1/6)\ [1\bar{1}\bar{2}]$$

and from Eq. (8.3-6)

$$\boldsymbol{x} = (1/6)\ [21\bar{1}]$$
$$\boldsymbol{y} = (1/6)\ [121].$$

Hence, $\operatorname{sgn}\left(\prod_i m_i\right) = (-)$ and $\operatorname{sgn}\left(\prod_i c_i\right) = (+)$ and thus $\operatorname{sgn}(\boldsymbol{B}\cdot\boldsymbol{m}) = (-)$ [Eq. (8.3-9)], i.e. the sense of $\boldsymbol{B}$ is opposed to that of $\boldsymbol{m}$. This condition is fulfilled if:

$$\boldsymbol{b}^{(1)} = \boldsymbol{y} = (1/6)\ [121]$$
$$\boldsymbol{b}^{(2)} = \boldsymbol{x} = (1/6)\ [21\bar{1}].$$

In order to check the invariance properties of Eq. (8.3-9), we assume $\boldsymbol{l}$, $\boldsymbol{b}$ and $\boldsymbol{v}$ as given (Fig. 8.3/6, case b) while $\boldsymbol{m}$, $\boldsymbol{c}$ and $\boldsymbol{B}$ follow from them. We introduce the following changes:

$$\boldsymbol{l} \to -\boldsymbol{l}, \qquad \boldsymbol{b} \to -\boldsymbol{b}, \qquad \boldsymbol{v} \to \boldsymbol{v}.$$

(Due to the pseudo-vector character of the Burgers vector this must leave the physical situation unchanged.) Here $\boldsymbol{m} \to -\boldsymbol{m}$ and thus $\operatorname{sgn}\left(\prod_i m_i\right) \to -\operatorname{sgn}\left(\prod_i m_i\right)$ while $\operatorname{sgn}\left(\prod_i c_i\right) = \operatorname{sgn}\left(\prod_i [\boldsymbol{b}\times\boldsymbol{m}]_i\right)$ remains unchanged as both $\boldsymbol{b}$ and $\boldsymbol{m}$ have changed their signs. Thus:

$$\operatorname{sgn}(B\cdot m) \to -\operatorname{sgn}(B\cdot m)$$

but, since $\boldsymbol{m}$ has changed, $\boldsymbol{B}$ stays unchanged.

From this, it follows that

$$\boldsymbol{b}^{(1)} \to -\boldsymbol{b}^{(1)} \quad \text{and} \quad \boldsymbol{b}^{(2)} \to -\boldsymbol{b}^{(2)}.$$

Since, at present, the $(-\boldsymbol{m})$-side is displaced, due to the movement of the dislocation, the triangle formed by $\boldsymbol{b}^{(1)}$, $\boldsymbol{b}^{(2)}$ and $\boldsymbol{b}$ is inverted.

Further physically equivalent cases are:

d) $\quad \boldsymbol{l} \to \boldsymbol{l}, \qquad \boldsymbol{b} \to \boldsymbol{b}, \qquad \boldsymbol{v} \to -\boldsymbol{v},$

e) $\quad \boldsymbol{l} \to -\boldsymbol{l}, \qquad \boldsymbol{b} \to -\boldsymbol{b}, \qquad \boldsymbol{v} \to -\boldsymbol{v}$

and physically opposed is

f) $\quad \boldsymbol{l} \to -\boldsymbol{l}, \qquad \boldsymbol{b} \to \boldsymbol{b}, \qquad \boldsymbol{v} \to \boldsymbol{v}.$

Case f can be compared with the others, for example by placing the arrangements f and e (Fig. 8.3/6) side by side and letting the partials annihilate each other in a stepwise manner.

In the physically equivalent cases b to e $\boldsymbol{v} \to -\boldsymbol{v}$ involves an exchange of the numbering of $\boldsymbol{b}^{(1)}$ and $\boldsymbol{b}^{(2)}$ [while $(\boldsymbol{l} \cdot \boldsymbol{b}) \to (-\boldsymbol{l} \cdot -\boldsymbol{b})$ describe identical situations]. The orientation of the Burgers vector triangle is coupled with the orientation of $\boldsymbol{m}$ and thus with $\boldsymbol{v}$ and $\boldsymbol{l}$.

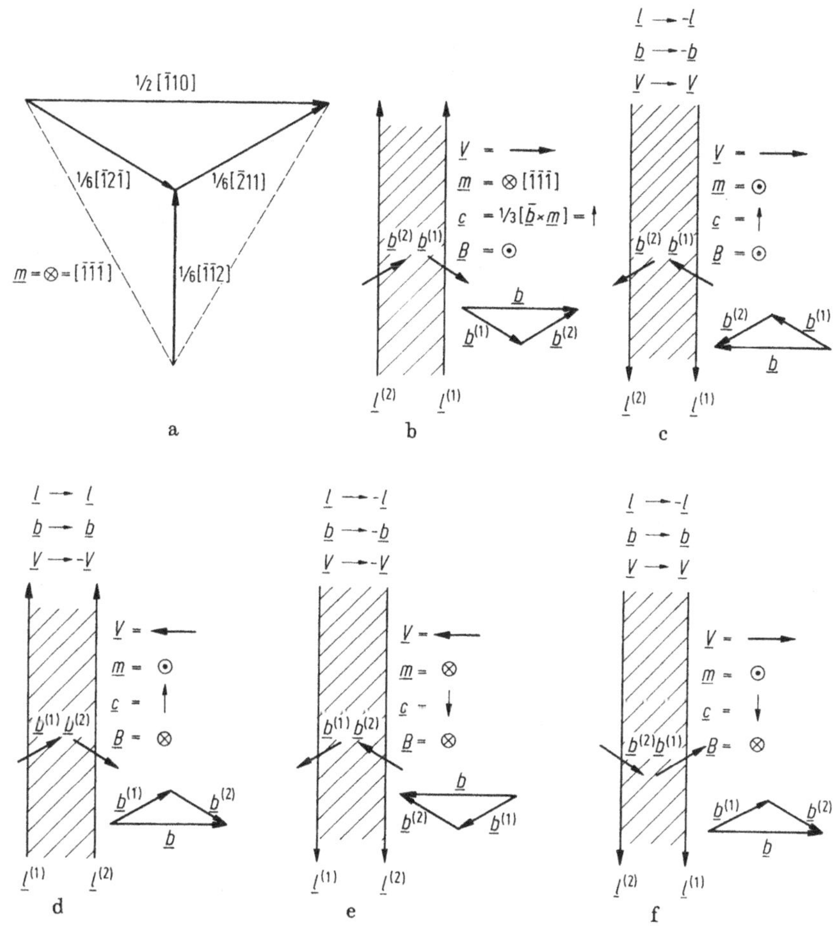

Fig. 8.3/6. Invariance properties of Eq. (8.3-13)

8.4. Transformation of the Face-Centred Cubic — into the Close-Packed Hexagonal Structure

Certain metals, a typical example being cobalt, transform on cooling from the fcc to the cph structure. Since the cobalt transformation occurs at around 420 °C, it cannot be diffusion-aided. The motive force for the transformation can be understood as being due to the decrease in the stacking fault energy, γ, with decreasing temperature; γ becomes zero at the transition temperature and negative below this temperature. This means that below 420 °C the H-sequence is perferred to the K-sequence.

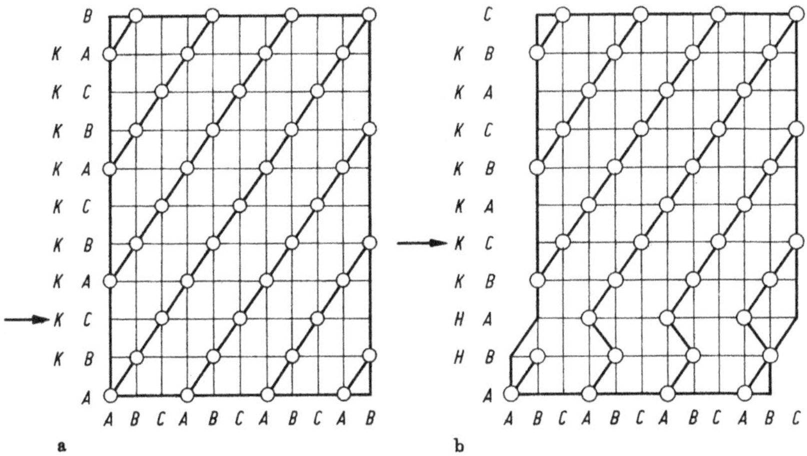

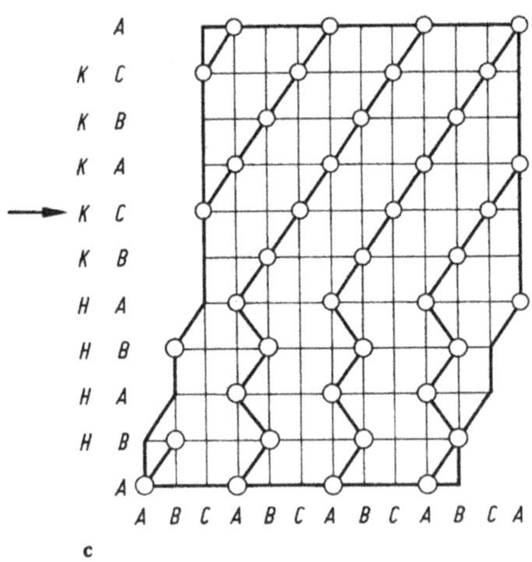

Fig. 8.4/1a–c. Geometrical conditions for the transformation of the f.c.c. into the c.p.h. structure. a), b), c) sequence of displacements

The transformation can be explained geometrically as a shift of every second layer of the K-sequence into the stacking fault position (Fig. 8.4/1):

$$\begin{array}{c} \downarrow \\ ABCABCABC \\ KKKKKKK \end{array} \qquad (8.4\text{-}1)$$

becomes

$$\begin{array}{c}\downarrow\\ ABAB\mathbf{CABCA}\\ \mathbf{HH}KKKKK\end{array} \tag{8.4-2}$$

and

$$\begin{array}{c}\downarrow\\ ABABAB\mathbf{CAB}\\ \mathbf{HHHH}KKK\end{array} \tag{8.4-3}$$

etc.

The question now arises as to how the transformation actually occurs. Electron microscopic observations show that the transformation is always incomplete. A survey of the problems can be found in Christian and Swann (1964).

We discuss here several possibilities for the transformation mechanism:

1. A closed loop of a partial dislocation, i.e. a circular disk of stacking fault, is formed spontaneously and starts to extend. This leads to a decrease in the energy of the crystal due to the increase in the stacking fault surface with negative stacking fault energy. On the other hand, the energy is increased by the extension of the partial dislocation. As the energy decreases $\propto r^2$ and increases $\propto r$, it will rise at the beginning, reach a maximum value, and will then decrease. The radius of the stacking fault disk at the energy maximum ($\delta E = 0$) is the critical radius, r_c. For $r > r_c$ the loop can extend spontaneously. r_c may be obtained from

$$\delta E = -\gamma \pi [(r + \delta r)^2 - r^2] + 2\pi \, \delta r \, \Gamma = 0 \tag{8.4-4}$$

with

γ = absolute value of the stacking fault energy,

Γ – line tension $\approx \frac{G b^2}{2}$.

This leads to

$$r_c \approx \frac{G b^2}{2} \tag{8.4-5}$$

with

$G \approx 7.5 \cdot 10^{11} \frac{\text{dyn}}{\text{cm}^2}$,

$b = \frac{a}{6} \langle 112 \rangle, \qquad b^2 = \frac{a^2}{6}$,

$a = 3.55$ Å,

$\gamma \approx 5 \frac{\text{erg}}{\text{cm}^2}$.

(In order to obtain a finite negative value for γ, the material has to be undercooled before start of the phase transformation. In fact, it usually starts at around 390 °C.)

With these values, r_c becomes $\approx 1{,}600$ Å. A stacking fault disk of $\approx 0.3\ \mu$ diameter should thus be formed spontaneously before it is able to grow on its own. The spontaneous formation of such a large loop is practically excluded.

2. The possibility of a "reflection mechanism" occurring at the crystal surface has been put forward in order to explain the formation of the stacking faults necessary for the transformation (see Christian and Swann, 1964). It was based on the following concept: Upon cooling, dislocations already existing in the fcc cobalt split into partials. The included stacking fault ribbon expands until it reaches the crystal surface where it produces a glide step. It was thought that, by dynamic effects new partial dislocations could be formed at the crystal surface in the surroundings of the glide step. These newly formed stacking faults could traverse the crystal in opposite directions and so trigger the formation of new stacking faults, and so on. The merit of this mechanism would be that the partial dislocation could be formed almost as a straight line and so would not have to overcome a critical radius of curvature. However, it is difficult to understand how new partial dislocations could be formed at the crystal surface, since the glide steps increase rather than decrease the surface energy. This is the reason why this concept has been abandoned.

3. A "pole mechanism" has been worked out by Seeger (1953, 1956). It is based on the assumption of the existence of a dislocation which is oriented essentially normal to a $\{111\}$-glide plane and which has a screw component of the type $(2/3)\langle 111\rangle$. A second dislocation of the type $(1/2)\langle 110\rangle$ branches off the first in the glide plane (Fig. 8.4/2). An example of such a combination is:

$$(1/2)\ [1\,2\,1] = (1/2)\ [\bar{1}\,1\,0] + (1/2)\ [2\,1\,1]. \tag{8.4-6}$$

The dislocation in the glide plane is split into two partials

$$(1/2)\ [\bar{1}\,1\,0] = (1/6)\ [2\,1\,1] + (1/6)\ [\bar{1}\,2\,1] \tag{8.4-7}$$

the first of which $(1/6)\ [2\,1\,1]$ moves in the sense of a right-handed screw in the $[1\,1\,1]$-direction and the second in the $[\bar{1}\,\bar{1}\,\bar{1}]$ direction. (This sequence can be confirmed as an application of the methods described in Section 8.3.) The effect of the assumed screw component is to displace every second layer by the Burgers vector of the partial dislocation, thus transforming the crystal from the cubic to the hexagonal structure.

The difficulties of this mechanism, as were already discussed by Seeger, are firstly that the two partial dislocations repel one another after half a revolution (Section 7.2), and that kinetic energy is necessary to surmount this barrier; this effect would explain the undercooling

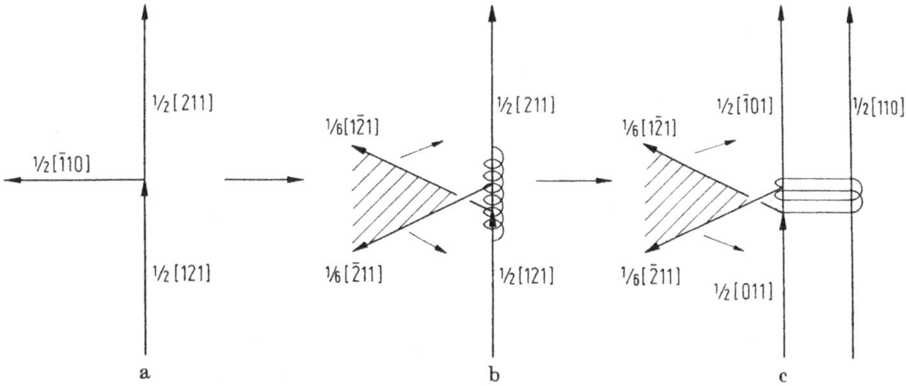

Fig. 8.4/2a–c. Seeger mechanism. a) Starting condition, b) pole mechanism: height of screw path = 2 layers, c) pole dislocation as two separate stable dislocations

which is necessary for the transformation. Secondly, it is obvious that the dislocations (1/2) [1 2 1] and (1/2) [2 1 1] are not stable:

$$(1/2)\ [1\,2\,1] = (1/2)\ [1\,1\,0] + (1/2)\ [0\,1\,1] \tag{8.4-8}$$

$$(1/2)\ [2\,1\,1] = (1/2)\ [1\,1\,0] + (1/2)\ [1\,0\,1]. \tag{8.4-9}$$

The dislocation combination necessary for the transformation would therefore not be formed at all, unless by chance a node of the type

$$(1/2)\ [0\,1\,1] = (1/2)\ [\bar{1}\,1\,0] + (1/2)\ [1\,0\,1] \tag{8.4-10}$$

was in the vicinity of a dislocation (1/2) [1 1 0] (Fig. 8.4/2c), with the partial dislocations winding around the free dislocation.

If the transformation mechanism were stopped, e.g. by other dislocations, stacking faults, etc., another dislocation arrangement of the same type would be necessary for the continuation of the transformation at a new location. Electron microscopic observation of a fully recrystallized material shows, however, that practically no dislocations are present inside the grains. This means that the dislocation combination required for the pole mechanism in cubic cobalt is extremely unlikely, although possible in principle.

4. As a further possibility, the "reflection mechanism" was reconsidered (Bollmann, 1961), with the modification that the new stacking faults were not formed at the crystal surface but internally, at differently oriented stacking faults which were assumed to be already present. Of course, also in this case, it is not a proper reflection, but rather the creation of a state of strain, which, in order to be relaxed, favours the

formation of new stacking faults moving in opposite directions (Fig. 8.4/3). The Burgers vectors of the new partials include an angle of 120° with that of the incoming ones. Hence, for example, two partials have to move out in order to cancel the long range strain resulting from the junction of the stacking faults. A mechanism of the type given in Fig. 8.4/4 results from these assumptions, whereby the different steps

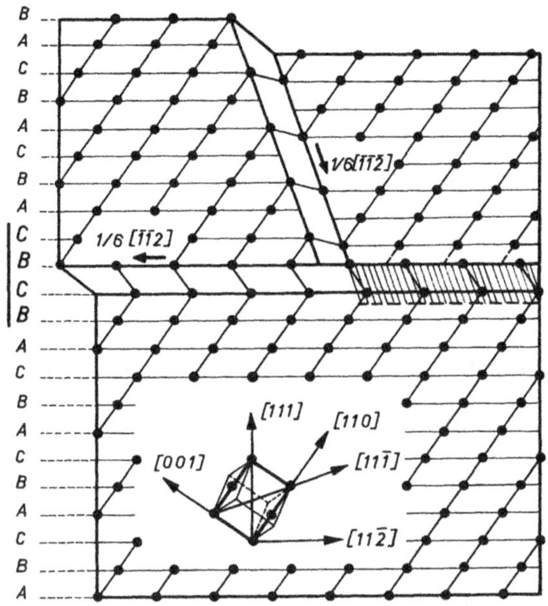

Fig. 8.4/3. Geometrical situation at the intersection of two stacking faults

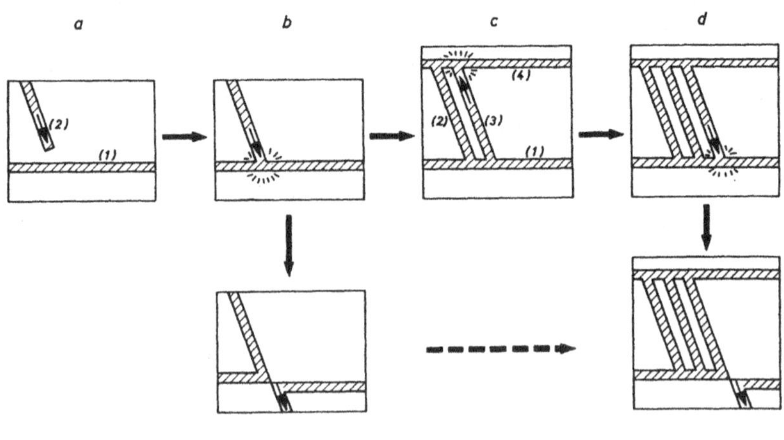

Fig. 8.4/4. Phase transition mechanism by "reflection" at stacking faults

are thought to take place on a statistical basis. Fig. 8.4/4a shows that at slow rates the stacking fault (2) would come to rest in an equilibrium position at a certain distance from the stacking fault (1). Only with the help of additional kinetic energy of the partial dislocation could new stacking faults be formed.

The essential points of such a mechanism are:

a) The basic configuration of converging stacking faults is frequent, since dislocations present in the cubic phase would tend to split. Remnants of former stacking faults can often be observed (Figs. 8.4/5, 8.4/6).

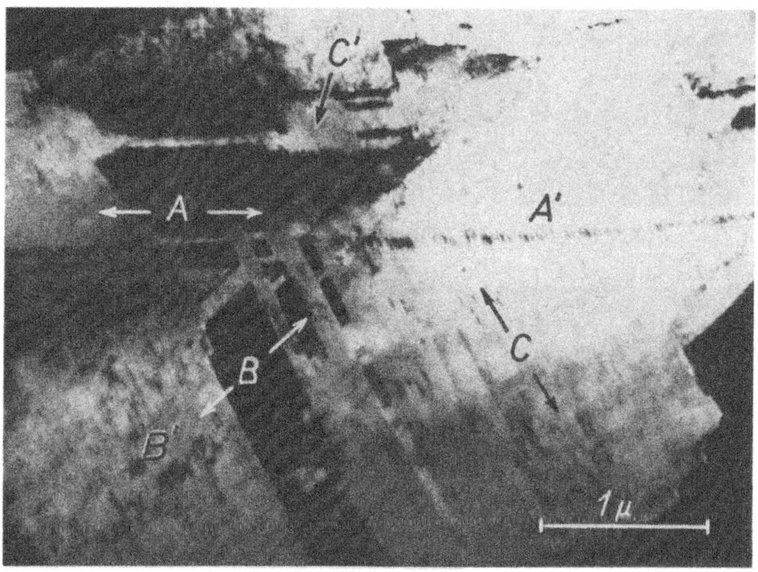

Fig. 8.4/5. Evidence that the final transformation in direction C took place across stacking faults already present in directions A and B

b) The newly formed partial dislocation is essentially straight since it had to be formed along the straight line junction of two stacking faults, without having had to surmount a critical radius (analogous to the "reflection at the surface", No. 2).

c) The transformation does not have to be continuous as is the case with the "pole mechanism". Many faults are possible and are also frequently observed (Figs. 8.4/5 and 8.4/6).

d) On slow cooling, a framework of stacking faults can stabilize the cubic phase (Fig. 8.4/7), especially if it is sufficiently narrow to prevent new partial dislocations from building up enough kinetic

Fig. 8.4/6. Probable remnants of former stacking faults

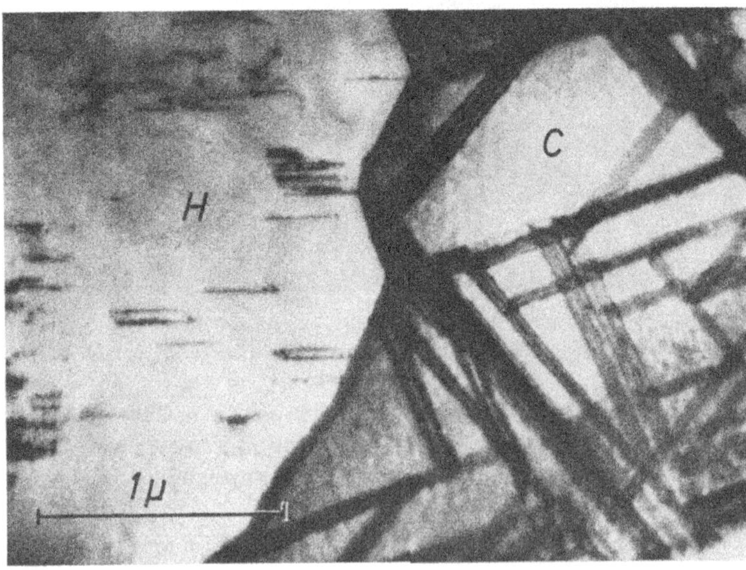

Fig. 8.4/7. Hexagonal grain (*H*) adjacent to a cubic grain (*C*) which is stabilized by a framework of stacking faults

energy to trigger the mechanism. On the other hand, on rapid cooling (quenching in liquid air), a more complete transformation is induced,

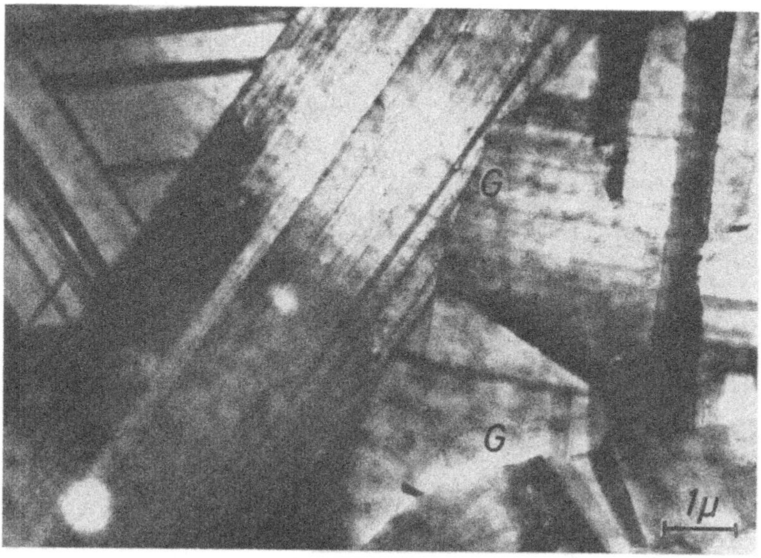

Fig. 8.4/8. Practically complete transformation induced by quenching in liquid air. *G* boundaries of the original cubic grain

whereby domains in the four different ⟨1 1 1⟩-orientations are produced in the original cubic grains (Fig. 8.4/8).

5. A comparable mechanism was suggested by Hirth (1969), in which it is assumed that an initially present stacking fault is cut by a perfect dislocation, so producing a step in the stacking fault, i.e. a partials dipole (Fig. 8.4/9). (A dislocation dipole is a pair of parallel dislocations of opposite sign.) Under the effect of the stress field, new dipoles would originate from the first and, as a result of the negative stacking fault energy, they would be forced apart, thus creating new layers of the hexagonal structure.

If, in the original assumption, a partial dislocation cuts the stacking fault instead of the perfect dislocation, the dipoles would have to be formed at a stacking fault. In this case the mechanism would be close to the previous one, except that the transformation proceeds parallel to the previously existing stacking fault, while in mechanism No. 4 the newly formed stacking faults are parallel to the incoming stacking fault.

6. Gedwill *et al.* (1964) observed that cobalt whiskers (hair-like single crystals) transform more readily in the longitudinal than in the transverse direction. The advantage from the energetic point of view becomes obvious if it is remembered that an oblique cut of a cylinder

is an elongated ellipse with the smallest radius of curvature at both extremities (Fig. 8.4/10). If the transformation starts at such an ex-

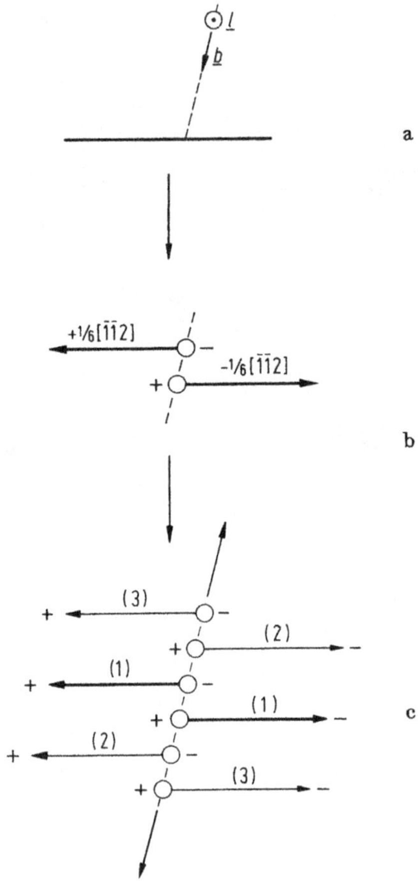

Fig. 8.4/9a–c. Dipole mechanism due to Hirth. a) Dislocation impinging on a stracking fault, b) dipole of partials formed at the limits of the "dislocated" stacking fault, c) formation of new partial dipoles. (The numbering sequence reflects the time sequence)

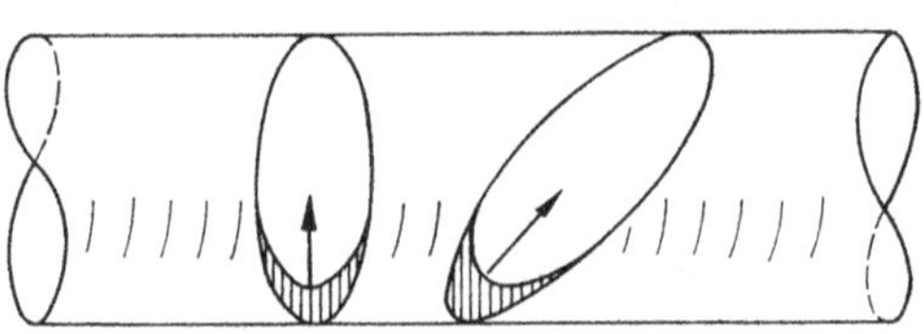

Fig. 8.4/10. Transformation of cobalt whiskers (schematic representation)

tremity, then the shear stress produced by the line curvature (Eq. (6.5-3)] and the increase in the extent of the stacking fault act together to accelerate the transformation. In contrast to case 1, the transformed part here lies outside the curved partial, and hence no critical radius has to be surmounted. The shear stress produced by the line curvature and the transformed surface per loop (i.e. the decrease in energy) are both larger in the case of the longitudinal transformation than for the transverse transformation. These considerations also hold when the shape of the whisker is prismatic instead of cylindrical. Thus, it is to be assumed that this kind of transformation nucleates at the surface.

9. Dualistic Representation of Dislocation Reactions

9.1. Introduction

In Section 7.4 it was shown that reactions take place between dislocations, i.e. they form nodes and coalesce if the energy of the configuration is thus lowered. At the nodes, Frank's node condition for the Burgers vectors (continuity of Burgers vectors) has to be fulfilled [Eq. (7.4-1)]:

$$\sum \boldsymbol{b}_{\text{in}} = \sum \boldsymbol{b}_{\text{out}}. \tag{9.1-1}$$

The Burgers vector is attributed to a dislocation but it is not localized on it. Here we choose a representation in which the Burgers vector is completely separated from the dislocation and placed into a separate space (Bollmann, 1962, 1964). The space containing the dislocation lines—the real crystal—is termed the L-space, and that of the Burgers vectors the b-space. While the L-space contains the dislocation lines together with their nodes, the b-space contains those Burgers vector configurations which arise from the fulfillment of the node condition. It will be shown that a strict duality exists between the two configurations. The situation is comparable to the force polygons of structural frameworks—the so called Cremona plans.

As a first example we discuss the intersection of two right-handed screw dislocations in the bcc lattice with the Burgers vectors (1/2) [1 1 1] and (1/2) [$\bar{1}\,\bar{1}$1] (Fig. 9.1/1 a). In Fig. 9.1/1 b the dislocation segments are numbered and shown together with their Burgers vectors. In Fig. 9.1/1 c, the line sense and the Burgers vectors of the segments 2 and 3 are inverted. Frank's node condition is in this case:

$$\sum \boldsymbol{b}_{\text{in}} = 0. \tag{9.1-2}$$

The Burgers vectors are now transferred to the b-space and added: since their sum is zero they must form a closed parallelogram configuration (Fig. 9.1/1 d). We shall now formulate a series of duality relations which will allow configurations in the L-space to be "translated" into those in the b-space and "vice versa".

The first relation, which we term D1, is as follows:

	L		***b***
D1	An *L-node*	corresponds to	a *closed b-polygon.*

This statement is obviously valid for arbitrary dislocations.

The coordinate system of the *b*-space is oriented in the same manner as that of the *L*-space, but for reasons of presentation, the units are usually chosen differently.

Instead of numbering the different dislocation segments we can also attribute numbers to the *L-fields* bounded by dislocations. We already introduce here a two-coordinate cipher system which will later be useful

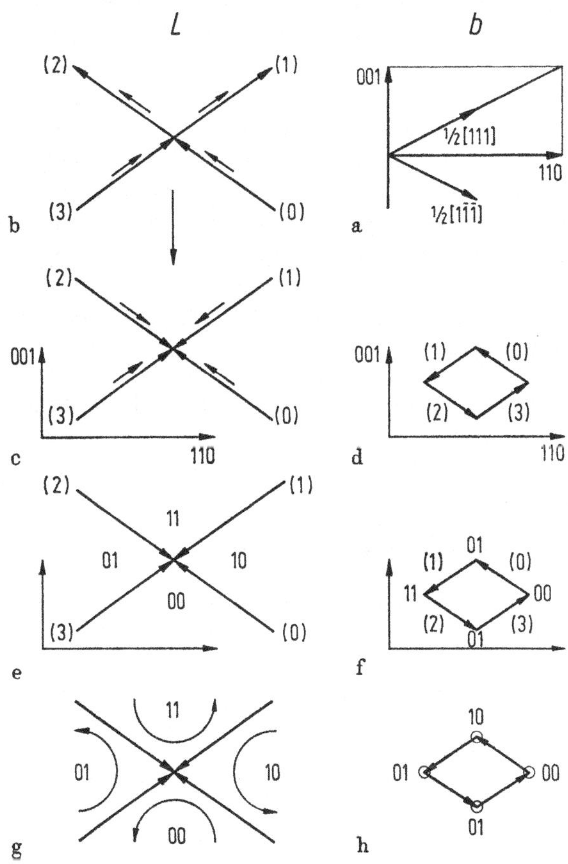

Fig. 9.1/1. Introduction to the dualities between the configurations of the dislocation lines and the Burgers vectors

for dislocation networks. The corresponding ciphers in the b-configuration are found at the *b-nodes* (Fig. 9.1/1 e, f). The number of Burgers vectors connected by a *b-node* equals the number of dislocation segments attributed to the corresponding L-field. Thus, D2 becomes:

D2	To every *L-field*	corresponds	a *b-node*.

Also, D3 is evident:

D3	A dislocation segment *separates* two L-fields.	The Burgers vector of that segment *connects* the two corresponding b-nodes.

Nothing has been said until now about the sense of the dislocation line and of the Burgers vector. If a dislocation segment bounds an L-field, its sense determines a sense of rotation around that field, which may be clockwise or counter-clockwise. In an analogous manner the Burgers vector may "flow" out of a node or into it, i.e. it can be fixed at the node either by its tail or by its tip. The relation between the sense of rotation and the flowing in or out of a node is remniscent of the relation between curl and divergency in electrodynamics. In order to settle a convention, we assume the principal sense of rotation around an L-field to be counter-clockwise. Thus, we see from Fig. 9.1/1.g, h that, if the dislocation segment follows this principal sense of rotation, the corresponding Burgers vector flows *out* of the corresponding node. Therefore we can formulate D4 as:

D4	If a dislocation segment flows in the principal sense *(counter-clockwise)* around an L-field.	then the corresponding Burgers vector flows *out* of the corresponding b-node.

and vice versa (see also Section 11.2).

An analogous relation exists between the sense of rotation around a b-polygon and the flow in or out of an L-node

D4′	If a dislocation flows *out* of an L-node.	then the corresponding Burgers vector points *clockwise* around the corresponding b-polygon.

It is to be noted that, in the L-space, only the sense of the dislocation line and not the orientation of the Burgers vector is of importance. On the other hand, in the b-space, only the action of "flowing in or out" counts and again not the orientation of the Burgers vector. A Burgers vector drawn in an L-configuration serves only as an aid for visualizing the edge- or screw character. When the L- and the b-configuration are drawn, the line sense and the sign of the Burgers vector can be neglected. They can always be reconstructed from the diagrams and from the numbering scheme.

9.2. The Significance of the b-Net

According to the movement rule [Eq. (5.4-1)], the Burgers vector can be interpreted as a displacement of that part of the crystal marked by $\boldsymbol{m}$ if the dislocation moves in the direction $\boldsymbol{v}$. Let us assume that, at the beginning, the crystal is free of dislocations. The field 00 is then extended over the whole area. The dislocations are now allowed to enter the crystal one after the other and to form the dislocation configuration step by step. In each freshly created field, one part of the crystal is shifted with respect to the other. D4 was chosen such that, with the FS/RH sign convention, the b-configuration represents a map of the relative displacements of that part *above* the corresponding L-field. The part of the crystal *below* the L-configuration is considered as being fixed. Hence, the vector $\boldsymbol{m}$ of the movement rule has always to point upwards; this is the case if $\boldsymbol{l}$ points in the direction of principal sense. In this way, $\boldsymbol{b}$ flows out of the node 00 and into the neighbouring node (e.g. 00 $\rightarrow$ 10, Fig. 9.1/1g, h). However, if $\boldsymbol{l}$ points in the opposite direction to the principal sense, then $\boldsymbol{m}$ points downwards. In order that the b-configuration describes the displacements of the upper part of the crystal, either $\boldsymbol{l}$ and $\boldsymbol{b}$ have to be inverted or, in other words, $\boldsymbol{b}$ has to be connected from its tip to the b-node 00, i.e. it has to flow into the node 00. Fig. 9.2/1 presents schematically the displacement of the upper part with respect to the lower part the latter being kept fixed.

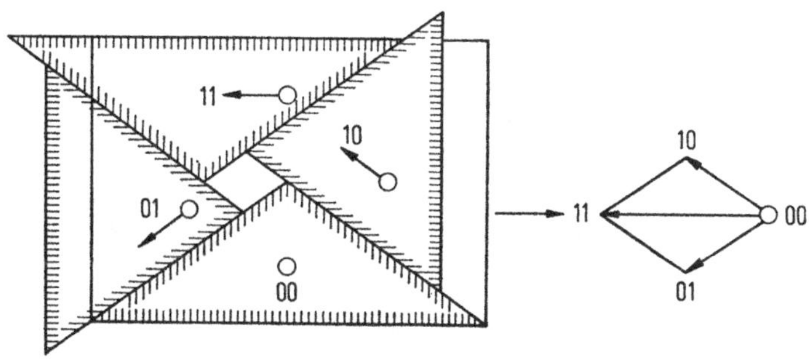

Fig. 9.2/1. Representation of the b-net as a map of the relative displacements

9.3. The Prediction of Dislocation Reactions

Based on the b^2-criterion we know from Section 7.4 that a dislocation reaction might take place if the Burgers vectors form an acute-angled triangle. We again consider our example from Section 9.1, with the two screw dislocations (1/2) [1 1 1] and (1/2) $[\bar{1}\bar{1}1]$, where the b-polygon forms a rhombus (Fig. 9.1/1 d). We divide this rhombus into two acute-angled triangles, and translate the configuration to the L-side. The b-nodes 1 0 and 0 1 are then directly connected, which according to D3, means that a direct separation (i.e. a dislocation) exists between the correspondingly numbered L-fields. This b-connection is the Burgers vector of the new dislocation line. The signs are attributed by D4. The two b-nodes (1 0 and 0 1) now join three Burgers vectors, which according to D1, D2 and D3, means that the corresponding L-fields are bounded by three dislocations (Fig. 9.3/1). The new dislocation is that reaction product which is energetically favored.

$$(1/2)\,[1\,1\,1] + (1/2)\,[\bar{1}\bar{1}1] = [0\,0\,1]$$
$$b^2\colon\ \frac{3a^2}{2} + \frac{3a^2}{2} > a^2. \qquad (9.3\text{-}1)$$

It is also oriented as a right-handed screw dislocation.

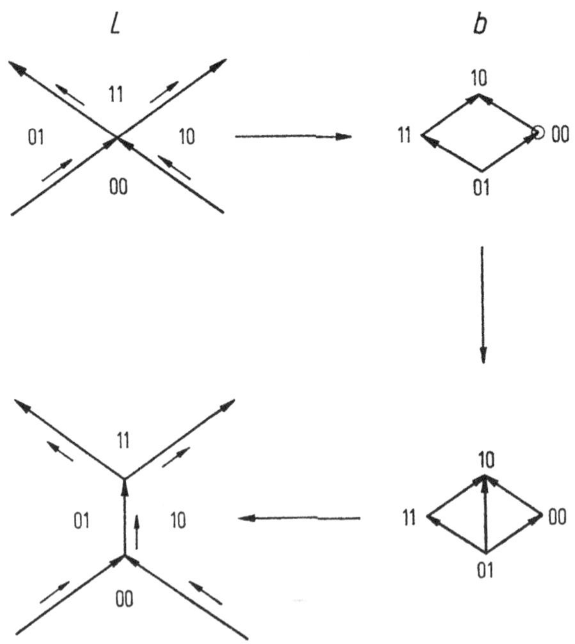

Fig. 9.3/1. Stable dislocation reaction

Hence, we can predict possible dislocation reactions by subdividing the b-configuration into acute-angled triangles and translating the new b-configuration to the L-side. On the L-side, naturally, only the topological connections, and not the exact geometrical L-configurations are obtained.

The subdivision of a b-configuration into obtuse-angled triangles would predict energy consuming reactions. The acute-angled or obtuse-angled triangle criteria are, strictly speaking, only meaningful if all the dislocations are of the same character. Fig. 9.3/2 presents an example in

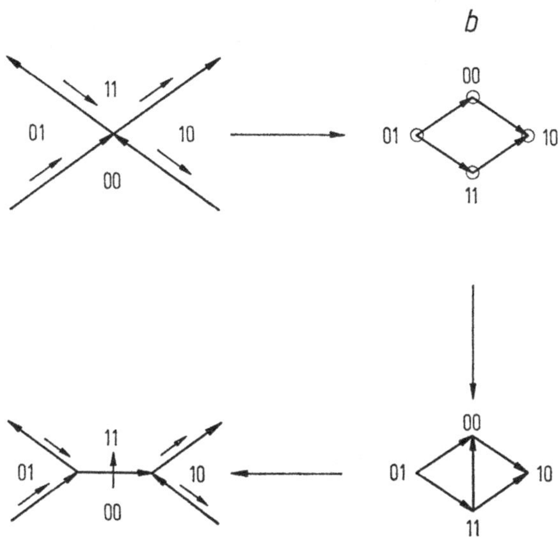

Fig. 9.3/2. Indifferent dislocation reaction

which edge and screw orientations are present. As the energy of an edge dislocation is 50% higher than that of the screw dislocation, it may happen in this case that the reactions do not occur, although the b-net has been subdivided into acute-angled triangles. It is to be noted that, in this case, a right-handed and a left-handed screw dislocation intersect. According to the numbering system in the L- and the b-configurations, it can be seen that the rotation sense is inverted in the two plots.

The energy of a dislocation is not only given by the Burgers vector and the edge and screw character, but also by its length. Dislocation reactions may also occur if energetically unfavorable Burgers vectors are produced, if the total energy can be lowered. In the bcc structure (iron) small segments of ⟨1 1 0⟩-dislocations have been observed as reaction products of two ⟨1 0 0⟩-dislocations (Hale and Brown, 1968).

In order to study a situation such as that given, for example, in Fig. 9.3/3a, the line sense of every dislocation segment is first fixed. It is advantageous to draw the Burgers vector corresponding to every line segment. Then the L-fields are numbered according to a two-co-

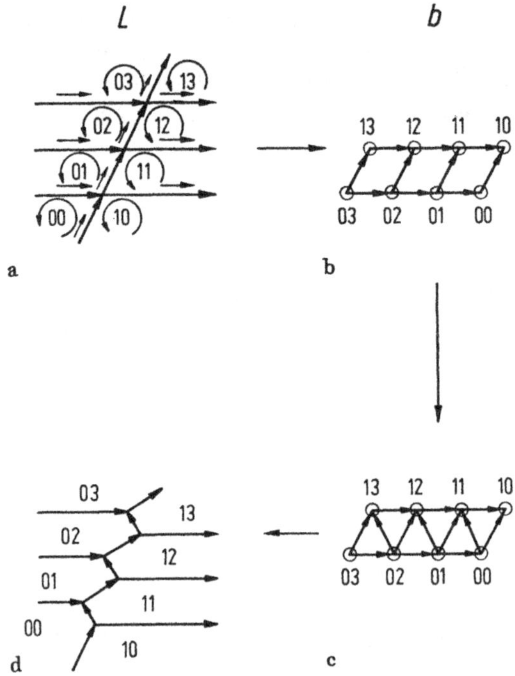

Fig. 9.3/3. Zig-zag reaction

ordinate system. Both, the dislocation lines and the Burgers vectors are referred to the same coordinate system. Now, on the b-side, a starting point (e.g. the point 00) is chosen, and according to D4, we move round the field 00. The corresponding Burgers vectors are accordingly joined, either by their tails or by their tips to the b-node 00. The other ends of the Burgers vectors are then numbered according to D3; every end receiving the number of the corresponding neighbouring field. Afterwards the construction continues in an analogous manner by moving round the neighbour fields one after the other, drawing the Burgers vectors and again numbering the ends, until every L-field is attributed to a b-node and the Burgers vector of every dislocation segment has found its place in the b-configuration. When all the fields have been dealt with in this way the whole b-configuration is constructed. Different projections may be required in complicated cases.

Once the b-configuration is drawn, it is subdivided, as far as possible, into acute-angled triangles and the new configuration is translated to the L-side. The number of Burgers vectors meeting in a b-node is equal to the number of dislocation segments bounding the corresponding L-field. Fig. 9.3/3 shows the so called zig-zag reaction.

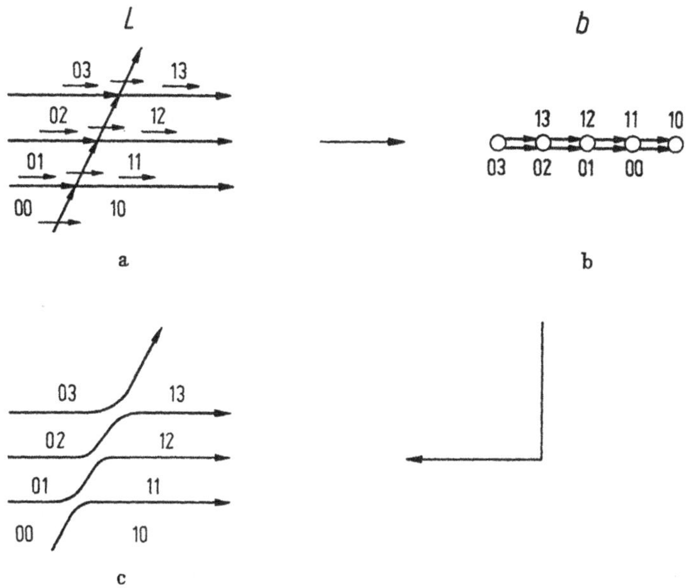

Fig. 9.3/4. Intersection of dislocations containing identical Burgers vectors

The example in Fig. 9.3/4 is a case where dislocations of different orientations, but with the same Burgers vector, intersect. The situation can be obtained from the foregoing example Fig. 9.3/3 by slowly turning the Burgers vector of the inclined dislocation line against the horizontal orientation. In this manner, certain b-nodes are merged so that now two numbers are attributed to the same b-node. As every b-node is, by definition (D2) attributed to a single L-field, the fact that a b-node is marked by two numbers means that two fields which were originally drawn separately become joined to one field, i.e. there is no displacement between these fields. We may formulate this as D5:

D5	Two L-fields, originally drawn separately, are joined to one single field ...	... if the numbers attributed to the L-fields meet at a single b-node.

The third example in Fig. 9.3/5 represents an *unstable* situation. This configuration can be stabilised in two ways by the formation of direct connections in the b-net; this corresponds to dislocation reactions in the L-net. An extension of $D2$ is evident from this example:

D6	A series of parallel dislocation lines	corresponds to	a chain of Burgers vectors connected in succession.

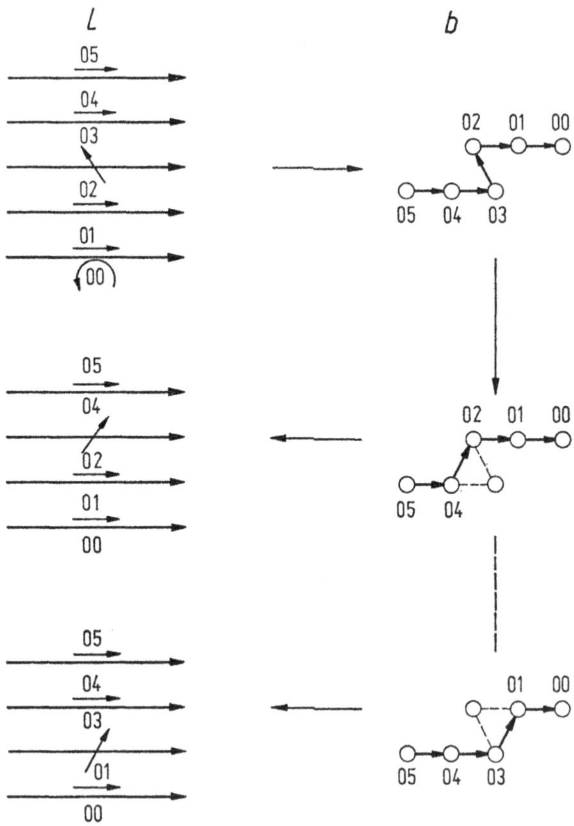

Fig. 9.3/5. Stabilization of an unstable dislocation arrangement

Since the dislocations are not joined at nodes, the b-polygon chain is not closed.

The procedures given in this section can also be extended to partial dislocations, as is done in the following section.

9.4. Dislocation Reactions Involving Partial Dislocations

In Section 8.3 we showed that, if a dislocation in the face-centred cubic structure (ABC-stacking) is split into partials, the sequence of the partials is fixed. In the hexagonal structure (the $ABAB$-stacking) both sequences can appear, according to whether the dislocation is located above an A- or a B-layer. Dislocation reactions can also take place if the dislocations are not on the same glide plane. In this case the dislocations

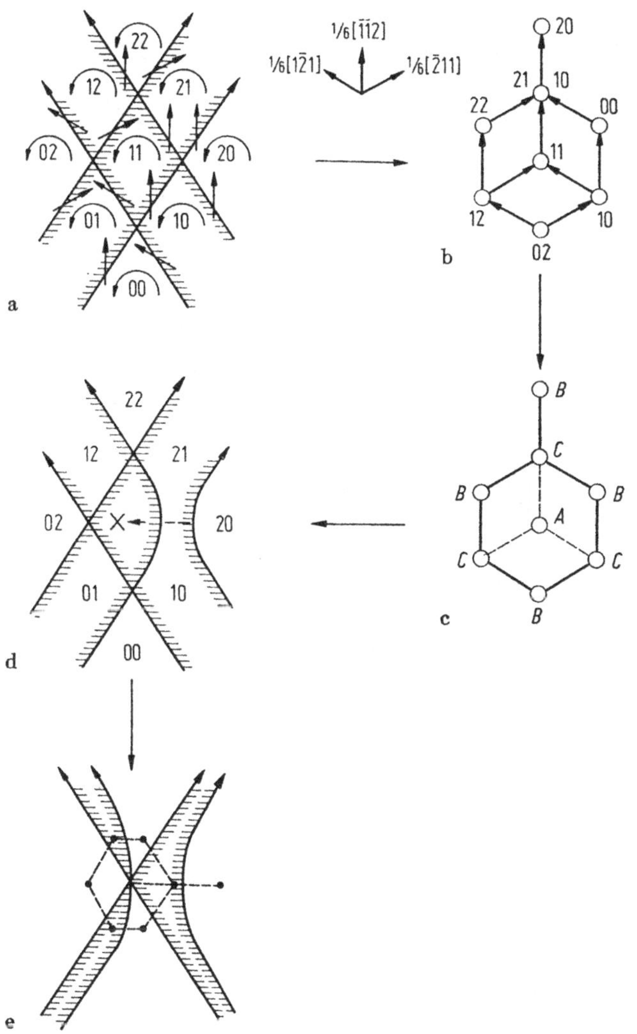

Fig. 9.4/1. Reaction of partial dislocations (triangular node)

align themselves parallel to one another instead of merging, and dislocations which would cancel each other constitute dislocation dipoles, i.e. a pair of parallel dislocations of opposite sign.

Here we investigate the crossing of two split dislocations. The first case (Fig. 9.4/1) can arise in the cubic as well as in the hexagonal structure. The dislocations may lie—but not necessarily so—in the same glide plane. When the perfect Burgers vectors and the glide plane are known,

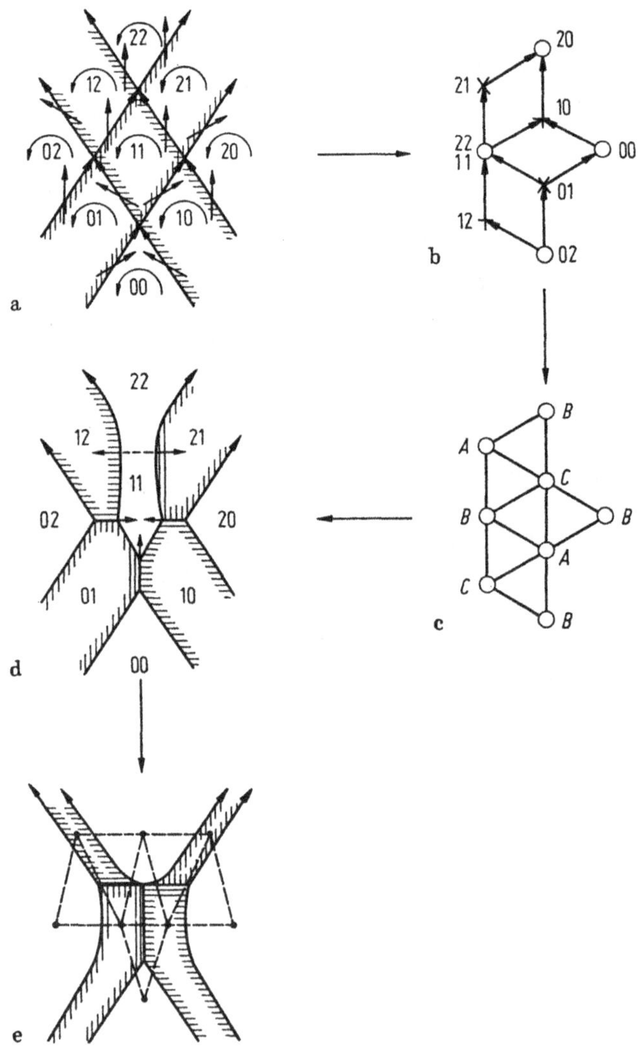

Fig. 9.4/2. Reaction of partials in the hexagonal structure

the sequence of partials has first to be determined according to Section 8.3. The Burgers vectors and the line sense of the partials are given for every dislocation line segment. Now, in order to construct the b-net, we number, and move round, the L-fields in the manner already described (Section 9.3). The b-net is a map of the relative displacements of the part of the crystal above the plane of the paper with respect to the lower part which is assumed to be fixed. Let us assume the first layer above the plane of the paper to be a B-layer and the structure below the plane of the paper to consist of a CA-stacking. As the upper layer can only sit in positions B and C (while A is forbidden since it is labile), the fields: 00, 20, 02, 22 have to remain in the B-position, and the fields: 10, 01, 12, 21 in the C-position i.e. the stacking fault position. It is seen in the b-net that the field 11 would remain in the A-position which is forbidden. Hence, this field has to contract to a point. Furthermore, the fields 21 and 10 have to join or to remain separated by a dislocation dipole, as they correspond to a common b-node (D5). Therefore, after retranslation, we obtain the L-configuration Fig. 9.4/1 e.

The second case (Fig. 9.4/2) can only occur in the hexagonal structure and then only if the two dislocations are in different positions (B above A and A above B). The configuration of this example is distinguished from that of Fig. 9.4/1 by the fact that in one of the dislocations the sequence of the partials is inverted. The b-net is again constructed in the usual manner. The fields: 00, 20, 02, 22 again contain no stacking fault, the fields 21 and 01 involve "C over A" stacking faults and the fields 10 and 12 "C over B" stacking faults. The fields 21 and 11 have to be joined. We subdivide the b-net into acute-angled triangles and retranslate the configuration. The partials 12/22 and 22/21 strongly repel one another, since the corresponding b-triangle represents the extreme case of an obtuse-angled triangle (180°). (See e.g. Hirth and Lothe, 1968, p. 327.)

Further reactions involving partial dislocations will be treated in Section 11.4.

9.5. Thompson's Notation of the Burgers Vectors

Thompson (1953) introduced the following notation in order to attribute a Burgers vector to every dislocation: As was shown in Section 8.3, the corners of Thompson's tetrahedron are denoted by the capitals A, B, C, D, and the centres of the opposite faces by the Greek letters α, β, γ, δ (Fig. 8.3/4). Two letters are attributed to every dislocation such that, on looking directly along the line (which fixes the line sense), the sequence of the letters is read, in the usual way, from left to right; for example, the letters AC denote the Burgers vector which, in the tetrahedron, points from A to C (Fig. 9.5/1). If the dislocation is viewed from the opposite side, the line sense, as well as the sequence of the letters, is inverted, which takes into account the pseudo-vector character of the Burgers vector. The perfect dislocations are denoted by

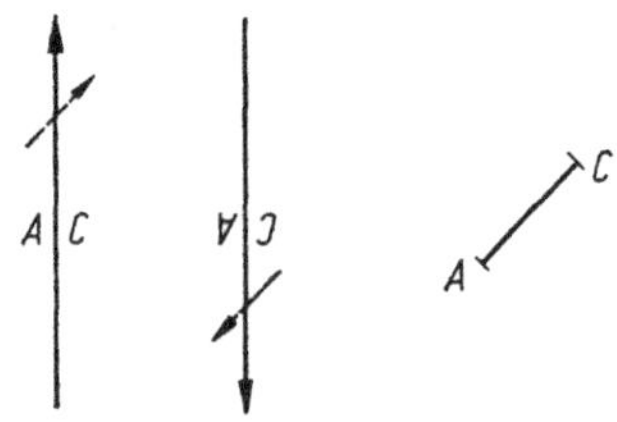

Fig. 9.5/1. Thompson's notation for a single dislocation

Roman letters, the partials by mixed Roman and Greek. A special "letter-algebra" was developed for the prediction of dislocation reactions, and which is widely used in the literature (e.g. Hirth and Lothe, 1968, p. 301, Amelinckx and Dekeyser, 1959). In Thompson's notation, the example Fig. 9.3/1 would correspond to Fig. 9.5/2.

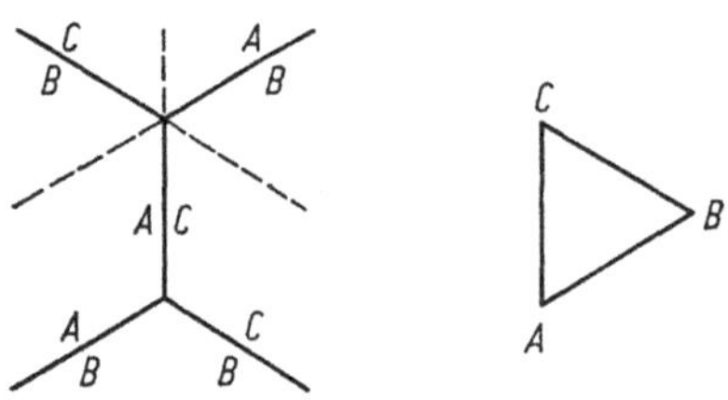

Fig. 9.5/2. Thompson's notation for two dislocation nodes

Thompson's notation is translated into the dualistic representation by replacing the letters in the dislocation segments by the Burgers vectors themselves, and by constructing the b-net according to Sections 9.3 and 9.4. If it is felt necessary, the letters can be reintroduced at the end. We shall restrict ourselves to the dualistic representation because, as a plan of the displacements between the L-fields, the b-net gives a clearer view of the situation than the letters in the L-net.

Chapters 7–9. Summary and Discussion

In Chapters 7 to 9, we started from the interaction between two parallel dislocations where we saw that dislocations of the same sign repel while those of opposite sign attract one another (Section 7.2). If only glide is allowed for parallel edge dislocations, then certain stable arrangements exist (Section 7.3).

Dislocations which attract one another can also merge partially or completely and form branches, i.e. they can react among themselves. Here, the important law of continuity of the Burgers vectors, i.e. Frank's node condition, has to be fulfilled (7.4). Besides this condition, the b^2-criterion (which, however, is only approximately valid) helps to decide whether a dislocation reaction might take place or not. Quantitative examples are given in Section 7.5.

Partial dislocations, representing partial displacements of a perfect dislocation, are treated in Section 8.2. When their configurations are considered they are seen to be limits of stacking faults. According to the b^2-criterion partial dislocations of a perfect dislocation repel each other, but they are held together by a strip of stacking fault, the width of which is determined by the stacking fault energy (surface tension).

The sequence of the partials in the fcc structure is not arbitrary, i.e. the addition of the partial Burgers vectors not commutative. The sequence can be determined geometrically by means of Thompson's tetrahedron and the movement rule [Eq. (5.4-1)], or algebraically by means of Eq. (8.3-13).

As an example of partial dislocations and stacking faults the phase transition of cobalt from the fcc- to the cph-structure is discussed in Section 8.4.

For the study of more complex dislocation reactions, the dualistic representation was developed, whereby the Burgers vectors and the dislocation lines are completely separated. Frank had already separated the Burgers vector from its dislocation by representing it in the reference lattice and not in the crystal. In principle, this idea is further developed here. On interpreting the node condition on a geometrical basis, a strict duality between the configurations of the dislocation lines and those of the Burgers vectors appears (Section 9.1). The b-net represents a plan of all the displacements on one side of the dislocation network with respect to the other side which is considered to be fixed (Section 9.2). The possibilities for dislocation reactions thus become directly visible (Section 9.3). Reactions of partial dislocations can also be treated in this manner (Section 9.4).

In order to interpret a given dislocation arrangement which has been observed by electron microscopy, for example, the following practice may help: A transparent paper is put on top of the photo. Then, in every field bounded by dislocations a representative point, if possible the central point of the field, is marked. Now all the neighbouring points are connected across the dislocation lines. Afterwards, the configuration is taken out of the dislocation network and rotated by 90°. It represents topologically, in most cases, the structure of the b-net although the sign is not determined. This procedure can be checked on the configurations of Sections 9.3 and 9.4 and the justification is given in the summary of chapter 11.

Suggestions for Problems

Many exercises can be found in the book "Elementary Dislocation Theory" by Johannes and Julia R. Weertman (1964) in which the subject of interactions between dislocations is more stressed than in the present book. Mathematically more sophisticated problems are treated in the book of Hirth and Lothe (1968).

The b^2-criterion can be more exactly formulated for dislocation reactions between dislocations of different character, whereby the limiting angle can deviate from 90°. In addition, other dislocation reactions such as are indicated in Section 7.5, can be discussed.

Data about the different possibilities of partials in the fcc and the bcc structures is to be found in the book of Weertman and Weertman (1964).

Further exercises on the dualistic representation of dislocation reactions can be set up by the combination of problems, as have been presented in Sections 9.3 and 9.4. It is, however, advantageous to relate these problems to dislocation networks (Chapter 11).

10. Short Description of the Relations between Dislocations and Point Defects, and Dislocation Dynamics

10.1. Dislocations and Point Defects

This section shall present a short description of notions which are important in dislocation theory, but which are not applied in this monograph. The details are found in the literature (e.g. Nabarro, 1967; Hirth and Lothe, 1968).

a) Dislocation loops can be formed by the condensation of vacancies or interstitials between densly packed layers. A supersaturation of vacancies may be produced in a material if it is quenched from a high temperature, at which thermal equilibrium existed. On slow heating, the vacancies become mobile and condense in disks or disappear at dislocations or grain boundaries where the edge components of the dislocations climb (Fig. 10.1/1). In a material with high stacking fault energy (e.g. aluminium) the condensed vacancies form loops of perfect dislocations while in a material with a low stacking fault energy (e.g. gold) small stacking fault tetrahedra are formed. Real holes, resulting from condensed vacancies, have been observed in Al-foils by Yoshida *et al.* (1963).

b) A distribution of point defects, not in the thermal equilibrium, can be produced by irradiation (neutrons, protons, electrons, fission products, γ-rays, etc.). Here also the point defects collect in groups upon heating, and the disturbed crystal lattice is thereby partially repaired. The different stages of recovery are strongly temperature-dependent. Problems of radiation damage are discussed in the book of Damask and Dienes (1963), for example.

c) When vacancies or interstitials condense on screw dislocations, helical dislocation lines may be formed.

d) Two notions which often appear in dislocation theory, especially in dislocation dynamics, are the "kink" and the "jog".

If a dislocation lies between two close-packed layers, then its energy will depend on the orientation (Fig. 3.2/1), whether it lies along rows of spheres (a configuration which is repeated every 60° in the $\langle 1\,1\,0 \rangle$-orientations) or in an arbitrary orientation. As a model, the glide plane can be represented as a row of parallel potential hills and valleys. A dis-

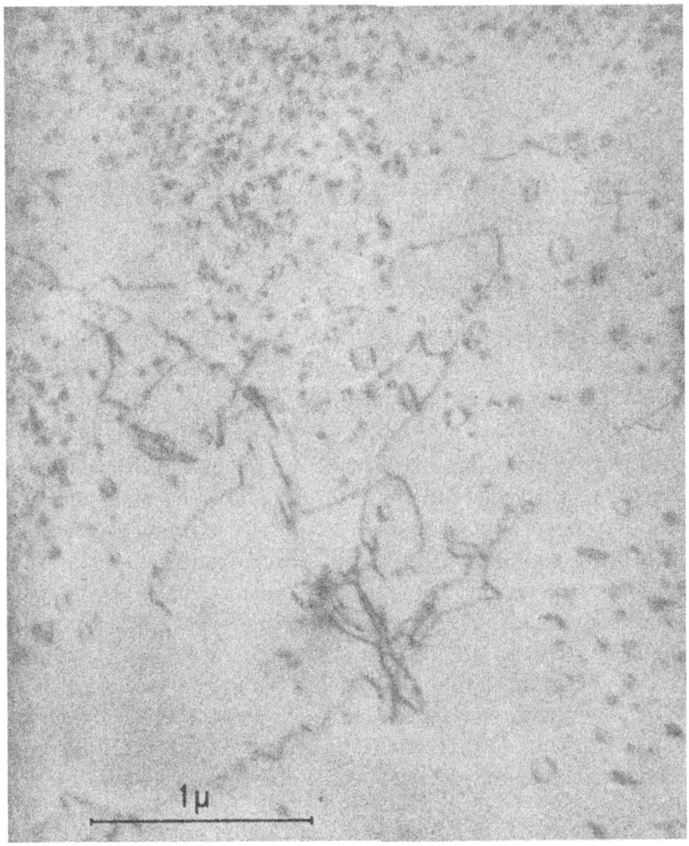

Fig. 10.1/1. Loops of condensed vacancies in aluminium. The irregular shape of the dislocation is due to climb of the edge components induced by condensed vacancies. (Photo: J. Silcox)

location will tend to align itself along the valleys. If it changes its path across a potential hill it will be bent (Fig. 10.1/2), and such local bending

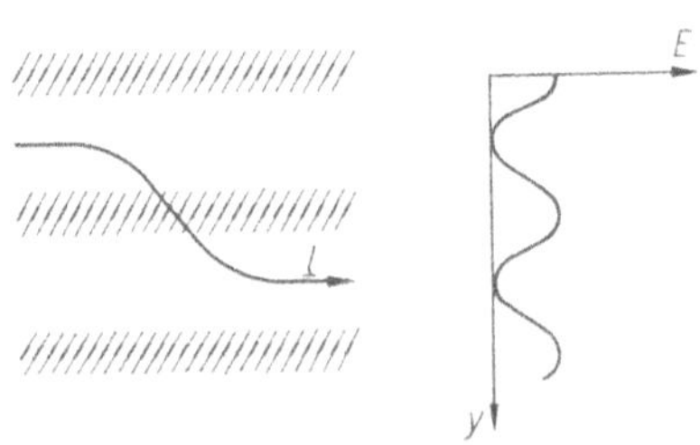

Fig. 10.1/2. Representation of a "kink"

is termed a "kink". When a dislocation moves in its glide plane, the kinks move along the line. This kink movement produces a friction force—the so called "Peierls-Nabarro force"—which impedes dislocation movement and contributes to the resistance to plastic deformation.

The other notion is the "jog". This is another case of a local bend in the dislocation line but this time with a component perpendicular to the glide plane (Fig. 10.1/3). A jog can be understood, for example, as a step in the extra plane of an edge dislocation. If, during the climb of an edge dislocation, the extra plane grows or diminishes, jogs move along the dislocation.

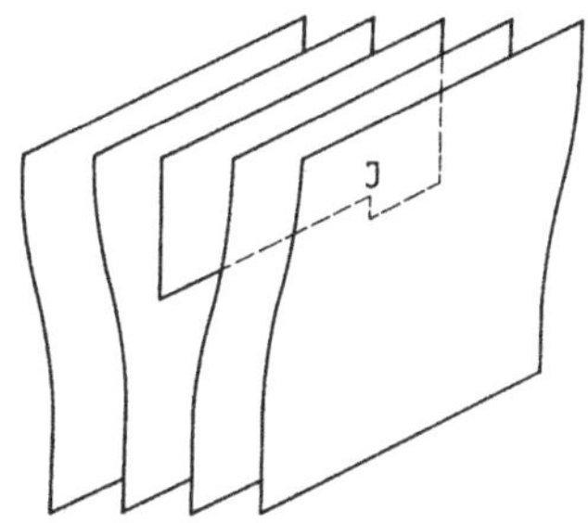

Fig. 10.1/3. Example of a "jog"

e) Kinks and jogs can be formed when two moving dislocations intersect. We investigate two right-handed screw dislocations which are normal to one another (Fig. 10.1/4). After the dislocations have crossed, a segment connecting the two lines is formed, in which the material is compressed. In this manner a chain of interstitials is formed, and energy is consumed. Thus, after intersection, the dislocations can no longer move freely. Each dislocation now has a jog which cannot glide, since the movement would involve a component perpendicular to the line segment of the jog as well as one perpendicular to the Burgers vector. If Fig. 10.1/4 is reconstructed for other examples, it is seen that:

1. The intersection of a right-handed and a left-handed screw dislocation produces vacancies, and

2. in the case of one or both dislocations being of pure edge character, a kink is formed without volume change, so that both dislocations can continue their paths relatively unimpeded.

If two dislocations intersect, dislocation No. 1 obtains a jog (or a kink) of the amount $\boldsymbol{b}_2$, and dislocation No. 2 one of $\boldsymbol{b}_1$. $\boldsymbol{r}$ shall be the direction of the movement of dislocation 1, where dislocation 2 is considered as fixed. Then, according to Hornstra (1962) (and Hirth and Lothe, 1968, p. 392), the sign of the displacement of the jog with respect

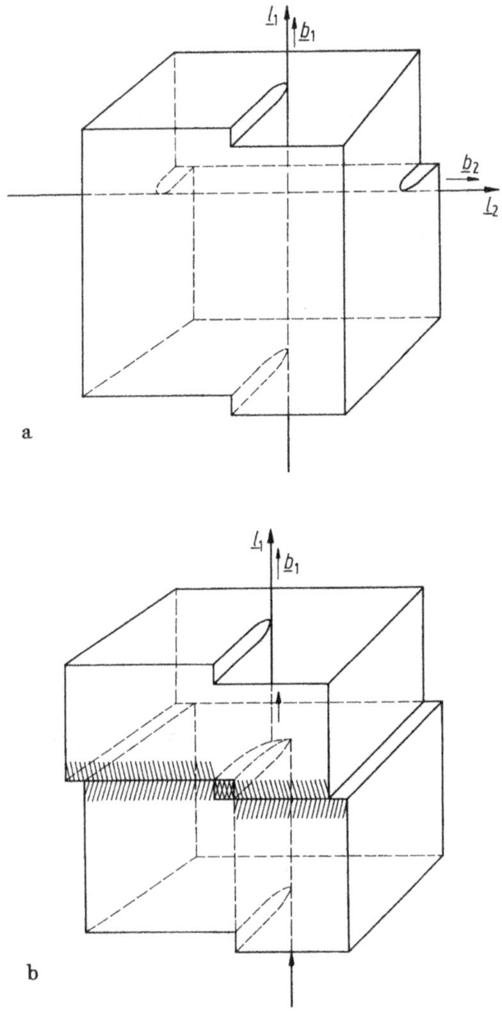

Fig. 10.1/4. Formation of interstitials by the intersection of two screw dislocations

to $\boldsymbol{l}_1$ is:

$$\operatorname{sgn}(\boldsymbol{b}_2) = \operatorname{sgn}\{\boldsymbol{l}_1 [\boldsymbol{r} \times \boldsymbol{l}_2]\}. \tag{10.1-1}$$

A jog in a screw dislocation acts as a local edge component with the corresponding extra plane. If the dislocation moves together with the jog, the extra plane of the jog either grows or diminishes, and vacancies or interstitials are thus produced. $\boldsymbol{r}'$ shall be simultaneously the direction of movement of the jog in dislocation 1 and the total amount of the displacement with respect to a coordinate system fixed in the crystal. The movement of the jog can have two components, one perpendicular to dislocation 1 (the same as $\boldsymbol{r}$) as though it were rigidly fixed to the dislocation, and the other along the dislocation. The number of point defects, N, produced on the path r' is, again according to Hornstra (1962):

$$N = \operatorname{sgn}(\boldsymbol{b}_2) \frac{\boldsymbol{b}_1 [\boldsymbol{r}' \times \boldsymbol{b}_2]}{V} \tag{10.1-2}$$

where $\operatorname{sgn}(\boldsymbol{b}_2)$ follows from Eq. (10.1-1).

$$\operatorname{sgn}(N) = (+) \text{ interstitials,}$$
$$= (-) \text{ vacancies,}$$
$$V = \text{atomic volume.}$$

If one or both dislocations are of pure edge character, $N = 0$.

10.2. Dislocation Dynamics

A large part of dislocation theory deals with dislocation dynamics. The papers of a symposium held on this subject are collected in the book "Dislocation Dynamics", edited by A. R. Rosenfield *et al.* (1968).

Dislocation dynamics treats the dynamical behaviour of dislocations during plastic deformation. One aim of studies in this field is the understanding of stress-strain curves. The relation between stress and strain depends on the velocity of deformation, on the temperature and on the history of the specimen. Even if it is not yet possible to give an exact quantitative explanation of these very complex phenomena, the study of dislocation dynamics contributes substantially to their understanding.

11. Dislocation Networks—Subgrain Boundaries

11.1. Introduction

In this chapter we shall investigate the case of two joined crystals of the same kind, but slightly rotated with respect to one another. In this context, the word "slightly" means that the two crystals are practically in register over areas which are large compared to the unit cell, and that only occasional corrections are needed in the form of dislocations. Such a boundary, consisting of a network of discrete dislocations, is termed a "subgrain boundary".

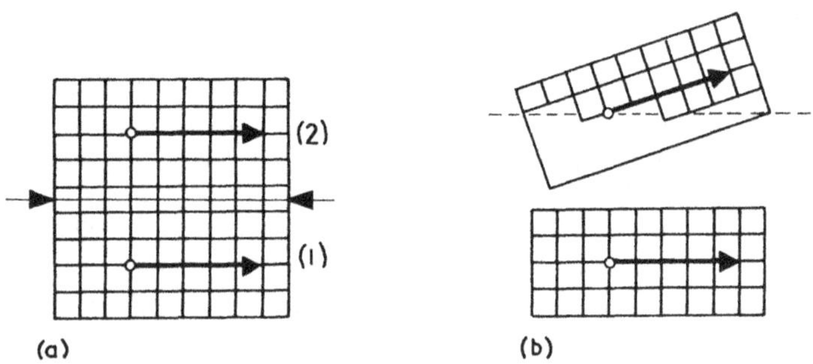

Fig. 11.1/1. Fitting of two crystals. a) Perfect crystal divided into two parts, b) both parts separated and crystal 2 rotated

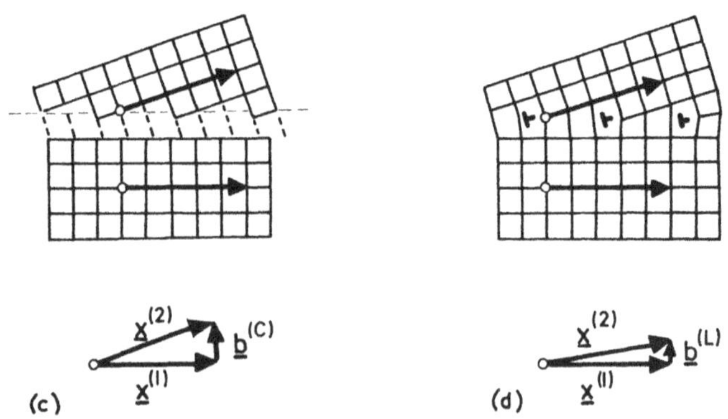

Fig. 11.1/1. c) Geometrical fit, d) physical fit

We visualize the formation of a subgrain boundary in the following manner: First we assume a perfect crystal divided into two parts: 1 und 2 (Fig. 11.1/1). In both parts, we assume a free vector $\boldsymbol{x}$, i.e. $\boldsymbol{x}^{(1)}$ and $\boldsymbol{x}^{(2)}$. In the perfect crystal both vectors shall be identical.

$$\boldsymbol{x}^{(2)} - \boldsymbol{x}^{(1)} = 0. \tag{11.1-1}$$

Instead of a single pair of vectors, we may imagine two vector fields in which every pair of related vectors fulfills condition (11.1-1).

Now we separate both parts and rotate crystal 2 by the (small) angle θ around the polar vector $\boldsymbol{\theta}$ ($|\boldsymbol{\theta}| = \theta$, orientation: right handed screw), leaving crystal 1 fixed. Along with crystal 2, the whole vector field is rotated, so that now the related vectors are no longer identical, i.e.

$$\boldsymbol{x}^{(2)} - \boldsymbol{x}^{(1)} = \boldsymbol{b}^{(C)}(\boldsymbol{\theta}, \boldsymbol{x}^{(1)}) \tag{11.1-2}$$

where, in the difference vector, $\boldsymbol{b}^{(C)}$, the superscript (C) stands for "continuous". We shall see that $\boldsymbol{b}^{(C)}$ is related to the Burgers vectors and that it is a function of the polar vector $\boldsymbol{\theta}$ and of the vector $\boldsymbol{x}^{(1)}$ (and respectively $\boldsymbol{x}^{(2)}$). We now cut the two crystals, while preserving the orientations, such that they can be placed on top of one another.

As long as we do not assume that atomic forces act between the two faces, the two superposed crystals form a moiré pattern at the interface (Moiré-Models Nos. 1–3). The locations where the crystal fit is perfect appear as bright areas, while those where there is misfit appear as dark areas. Such a situation is termed "geometrical fit".

If atomic forces act between the two crystals, then the bright regions of perfect fit will extend, and the dark bands, corresponding to misfit, will contract to dislocation lines, thus forming a dislocation network at the interface. This situation is called "physical fit".

In the next section the function $\boldsymbol{b}^{(C)}(\boldsymbol{\theta}, \boldsymbol{x}^{(1)})$ is investigated.

11.2. Frank's Formula

The function $\boldsymbol{b}^{(C)}(\boldsymbol{\theta}, \boldsymbol{x}^{(1)})$ as indicated by F. C. Frank (1950) can be written as:

$$\boldsymbol{b}^{(C)} = [\boldsymbol{\theta} \times \boldsymbol{x}] \tag{11.2-1}$$

with

$$\boldsymbol{x} \approx \boldsymbol{x}^{(1)} \approx \boldsymbol{x}^{(2)}. \tag{11.2-2}$$

Because of the small angle θ we do not distinguish between $\boldsymbol{x}^{(1)}$ and $\boldsymbol{x}^{(2)}$. In the form (11.2-1), Frank's formula is a purely geometrical statement. If $\boldsymbol{x} \perp \boldsymbol{\theta}$, then $|\boldsymbol{b}^{(C)}| = |\boldsymbol{x}|/\theta$. The important point, however, is the meaning which Frank attributes to $\boldsymbol{b}^{(C)}$. $\boldsymbol{b}^{(C)}$ is approximately equal to the sum of the Burgers vectors of the dislocations crossed by $\boldsymbol{x}$ if $\boldsymbol{x}$ is chosen in the boundary. We may illustrate this, for example, with the aid of the moiré model No. 1 (Fig. 11.2/1). The relation is only approximate because the Burgers vectors of dislocations can only acquire discrete values, while $\boldsymbol{x}$ and consequently $\boldsymbol{b}^{(C)}$ are continuously variable vectors.

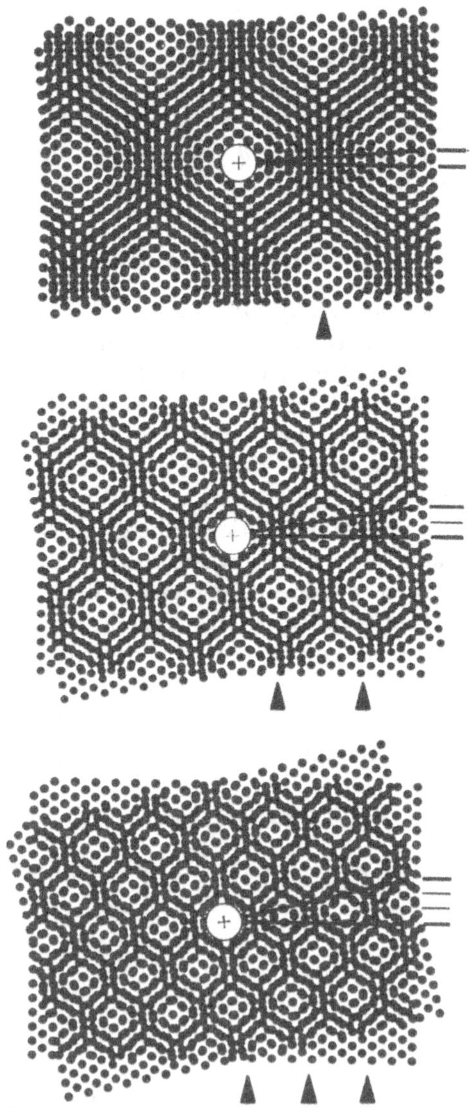

Fig. 11.2/1. Demonstration of Frank's formula ($b^{(C)} \approx n \cdot b^{(D)}$, n number of intersected dislocations)

According to Eq. (11.2-1), $\boldsymbol{b}^{(C)}$ is perpendicular to $\boldsymbol{x}$ as well as to $\boldsymbol{\theta}$. Frank's formula is an approximation. It is mathematically exact only for $\theta=0$, but in this case it loses its meaning. Also, it is obvious that for larger angles θ, $\boldsymbol{b}^{(C)}$ is always perpendicular to $\boldsymbol{\theta}$, but then $\boldsymbol{x}^{(1)}$ and $\boldsymbol{x}^{(2)}$ can no longer be identified. The generalisation of Frank's formula will be discussed in Section 12.3. For the time being, the theory of sub-grain boundaries will be developed on the basis of the approximation [Eq. (11.2-1)].

Referring to what was mentioned above, we may write Frank's formula as:

$$\sum \boldsymbol{b}^{(D)} \approx \boldsymbol{b}^{(C)} = [\boldsymbol{\theta} \times \boldsymbol{x}], \qquad (11.2\text{-}3)$$

where $\boldsymbol{b}^{(D)}$ is equal to the Burgers vectors of those *discrete* dislocations which are intersected by $\boldsymbol{x}$.

The Burgers vectors are translation vectors of the crystal lattice (or better: coordinate differences). As Frank's formula does not distinguish between the two lattices, the notion of the Burgers vector remains, in this context, somewhat indeterminate. It will be further clarified in Section 14.2.

We may interpret Eq. (11.2-1) as an imaging relation between the crystal space (x-space) and a b-space, where the correspondance between the two spaces is continuous.

Let us discuss this imaging relation more closely. We place $\boldsymbol{\theta}$ in an orthogonal coordinate system along the positive z-axis. Then Frank's formula, written in coordinate terms, is:

$$\begin{aligned} b^1 &= -\theta x^2 \\ b^2 &= \theta x^1 \\ b^3 &\equiv 0. \end{aligned} \qquad (11.2\text{-}4)$$

$b^3 \equiv 0$ signifies that only those points of the b-space which lie in the (b^1, b^2)-plane have an image in the x-space. This image is determined exclusively by x^1 and x^2 and is independent of x^3, which means that every b-point in the (b^1, b^2)-plane has, as its image, a straight line parallel to the x^3-axis, i.e. parallel to $\boldsymbol{\theta}$.

According to Eq. (11.2-3) there exist points in the (b^1, b^2)-plane for which:

$$\sum \boldsymbol{b}^{(D)} = \boldsymbol{b}^{(C)} = [\boldsymbol{\theta} \times \boldsymbol{x}]. \qquad (11.2\text{-}5)$$

To every one of these discrete points corresponds an individual straight line parallel to the axis of rotation in the x-space. In this manner, Frank's formula becomes "quantized", i.e. it is defined only for discrete b-points.

The x-lines can be separated from one another by *cell walls*. The cell walls shall be of the Wigner-Seitz type (Section 2.2), i.e. planes perpendicularly bisecting the connection between two x-lines. In this way a honeycomb type of cell structure is introduced into the crystal space.

We can interpret this construction as follows: Instead of being cut and joined, the two crystals can be idealized as two interpenetrating point lattices without space filling atoms. The interpenetrating lattices are understood as being two sets of positions for one set of atoms. The atoms can sit in either lattice and so "realize" the corresponding crystal.

We may now choose a boundary passing through the cell structure, and then place the atoms on one side of the boundary into lattice 1, and on the other into lattice 2. The intersection of the boundary with the cell walls becomes the dislocation network. The net, which is formed by the interconnections of those discrete b-points in the (b^1, b^2)-plane, the images of which (the x-lines) are intersected by the boundary, becomes the b-net. The relations between the two configurations are given by the dualities in Section 9.2. As soon as the boundary is fixed, the rest of the cell structure loses its meaning. Outside the boundary the space is now occupied by the undisturbed crystals 1 and 2. The intersections of the x-lines with the boundary are the points where the two crystals exhibit optimum matching. They are the centres of the fields bounded by dislocations (bright areas in the moiré model) and are

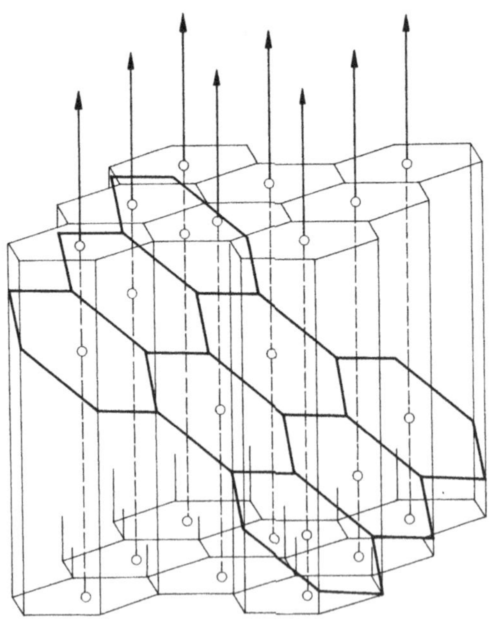

Fig. 11.2/2. Honeycomb structure representing the geometrical interpretation of Frank's formula (o = MS-points)

termed "minimum strain points", or "MS-points". The cell walls correspond to the locations of worst fit, i.e. the dark bands in the moiré models or the dislocations in the crystal (Fig. 11.2/2).

The b-net formed by all the Burgers vectors is a net of translation vectors of the crystal, which means it is a section through the translation lattice of the crystal. Hence, a natural coordinate system of the b-net, i.e. of the b-nodes, is given by the coordinate system of the crystal. In this manner, the corresponding "numbering" of the L-fields is also defined. In the configurations given in Section 9.1 the coordinates of the L-fields were first arbitrarily fixed, and those of the b-nodes follow.

While in earlier sections the L- and the b-configurations were related exclusively by geometrical-topological considerations, Frank's formula gives an algebraic relation between the two configurations of a subgrain boundary. For a given rotation $\boldsymbol{\theta}$, the cell structure is a three-dimensional representation of all the possible two-dimensional subgrain boundaries. Naturally, the cell structure is independent of the chosen coordinate system. The honeycomb-like cells are always parallel to the axis of rotation and the plane of all the b-vectors is perpendicular to that axis.

With respect to the application of the duality D4 (Section 9.1) a certain ambiguity remains in the attribution of the sign of the Burgers vector in relation to the dislocation line sense. If we define lattice 1 as the basis from which the rotation $\boldsymbol{\theta}$ starts and lattice 2 as the end product of the rotation, and if we now choose a boundary within the honeycomb structure, crystal 1 can be located on either side of the boundary. However, the sign of the dislocations in the boundary depends on the location of crystal 1. If crystal 1 and 2 exchange places, the dislocations change their sign (e.g. from a right-handed to a left-handed screw).

In order to take this fact into account the duality D4 can be reformulated in terms of the duality between curl and divergency, where the curl-vector is a polar vector (right-handed screw) coupled to the sense of rotation around the L-field (line sense), and the divergency is positive if the Burgers vectors "flow" out of the node (and vice versa). With these notions D4 becomes:

D4*	If the *curl vector of* an L-field *points into crystal 2*	the *divergency* of the corresponding b-node is *positive*.

Hence, if crystal 1 and 2 exchange sides the sign of the curl vector switches, i.e. the line sense alters while the Burgers vector remains unchanged which means that the dislocations switch their sign too.

In the next section a classification of subgrain boundaries is developed, based on this cell structure.

11.3. Classification of Subgrain Boundaries

A subgrain boundary is defined by:

a) the orientation of the rotation vector $\boldsymbol{\theta}$,

b) the value of the rotation angle θ and

c) the position of the boundary within the crystal (for the time being we assume a planar boundary).

a) We distinguish between a "special" and a "general" orientation of $\boldsymbol{\theta}$ within the crystal. As "special" we understand an axis of rotation which is (in crystal 1) perpendicular to two linearly independent Burgers vectors (minimal translation vectors). In this case the b-plane is densely occupied by b-lattice points which thus form a planar b-net. Such special orientations in the fcc structure are the $\langle 111 \rangle$- and the $\langle 100 \rangle$- and in the bcc case the $\langle 110 \rangle$- and the $\langle 100 \rangle$-orientations. In the "general" orientations of $\boldsymbol{\theta}$ it is no longer possible to construct a planar b-net in the b-plane which is defined by Frank's formula. This problem will be discussed in Section 11.6.

b) The value of θ determines the size of the cells; the larger θ, the smaller are the cells.

c) The orientation of the boundary plane with respect to the axis of rotation may be:

1. perpendicular,
2. in arbitrary orientation,
3. parallel.

Accordingly, the boundary is called:

1. a pure twist boundary,
2. a partial twist boundary,
3. a tilt boundary.

The terms are summarized in the following table:

Boundary	Orientation of $\boldsymbol{\theta}$	
	special	general
$\perp \boldsymbol{\theta}$	special pure twist boundary	general pure twist boundary
Arbitrary with respect to $\boldsymbol{\theta}$	special partial twist boundary	general partial twist boundary
$\parallel \boldsymbol{\theta}$*	special tilt boundary	general tilt boundary

* For the tilt boundary finer distinctions are needed, which will be given in Section 11.4.

11.4. "Special" Subgrain Boundaries

a) Special Pure Twist Boundary

In a special pure twist boundary the b-net is planar, and the b-net and the boundary plane are parallel. From Eqs. (11.2-4) and (11.2-5), follows that the b-net and the lattice of the MS-points are related by a

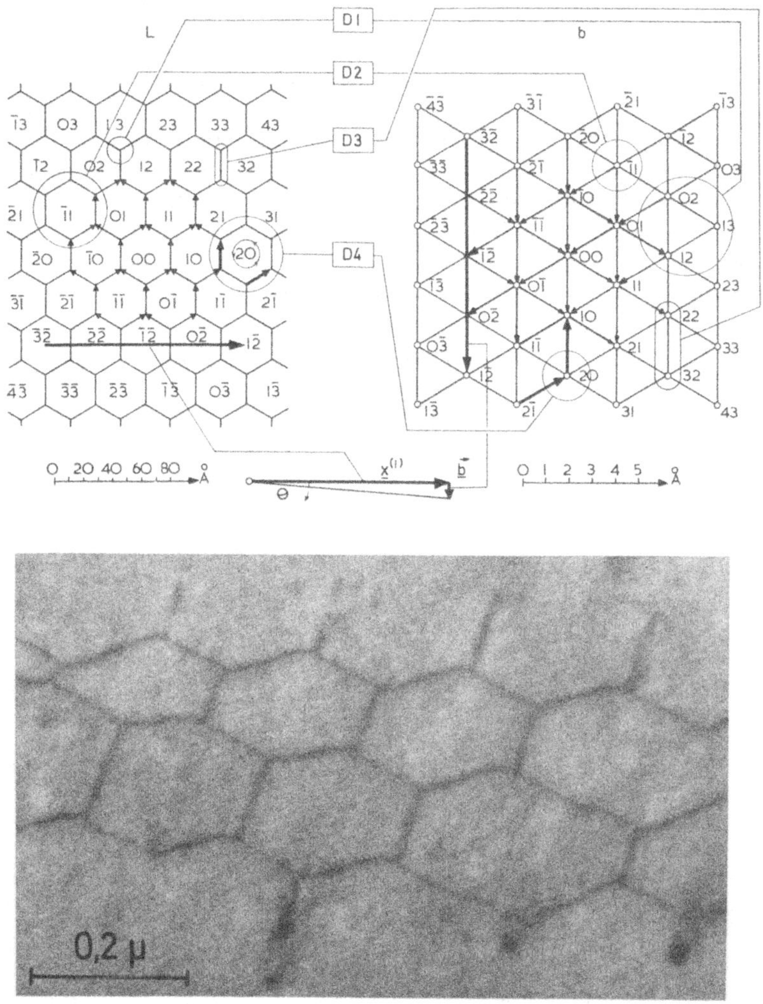

Fig. 11.4/1. a) Special pure twist boundary on the {1 1 1}-plane in the f.c.c. structure; duality between the L- and the b-net. b) Hexagonal dislocation network in iron. (Photo: A. Mascanzoni)

transformation in which the lattice of the MS-points is rotated by 90° with respect to the b-net. From this fact follows that the L-net consists of screw dislocations. Fig. 11.4/1 shows the duality relations between the b-net and the L-net for a hexagonal dislocation network in the $\{111\}$-plane of the fcc structure (or the basal plane of the cph structure). The corresponding networks in the $\{100\}$-planes of the fcc- and the bcc structures consist of squares. The $\{110\}$-plane in the bcc structure furnishes a somewhat distorted hexagonal network (moiré model No. 1).

b) Special Pure Twist Boundary of Partial Dislocations

Boundaries containing partial dislocations are possible in the fcc structure on $\{111\}$-planes, and in the cph structure on the (001)-plane, since stacking faults can be formed only on these planes. On splitting of perfect Burgers vectors into partial ones, the b-net, which was originally composed of triangles, is transformed into a hexagonal network (Fig. 8.3/1). Upon reverse-translation, an L-net consisting of triangles is generated in which every second triangle contains a stacking fault (Fig. 11.4/2). The single dislocation reaction corresponds to Fig. 9.4/1. An example in graphite is given in Fig. 11.4/3.

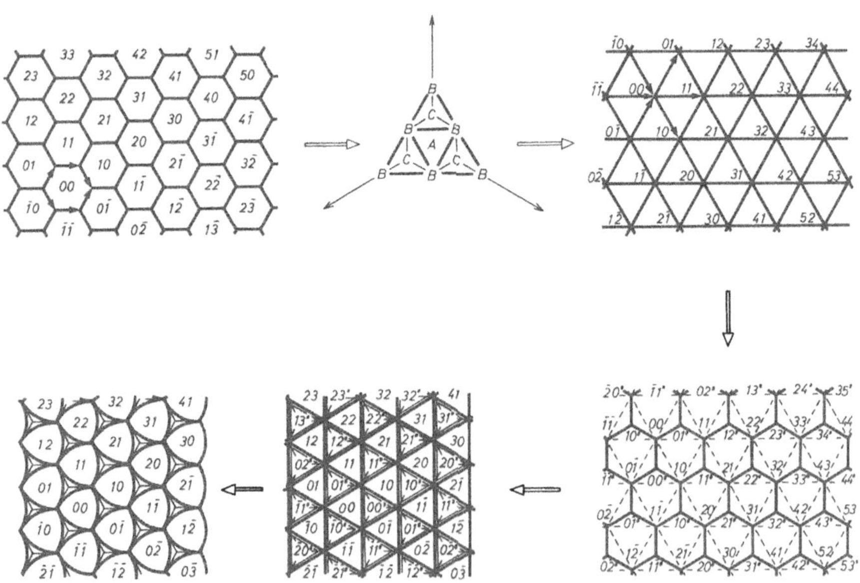

Fig. 11.4/2. Dislocation network of partial dislocations

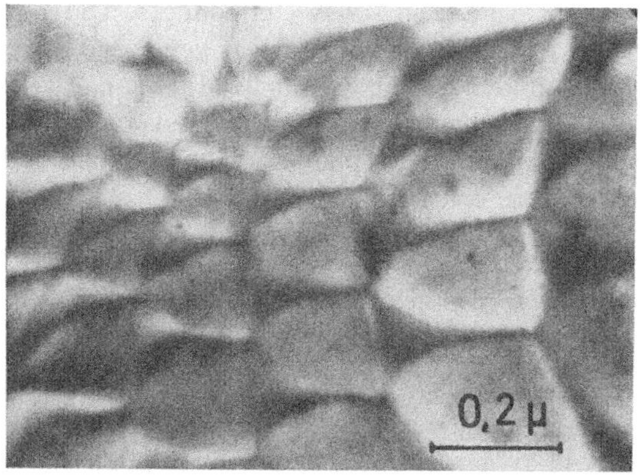

Fig. 11.4/3. Example of a triangular network in graphite

c) Special Partial Twist Boundary

In the case of the special partial twist boundary the same cell structure as for the pure boundary is cut by the now inclined boundary plane. Since the b-net remains unchanged, some of the dislocations acquire a

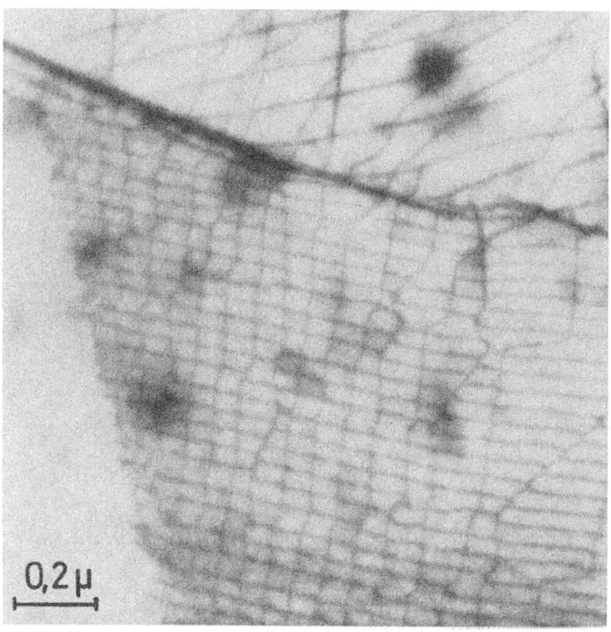

Fig. 11.4/4. Bent partial twist boundary. (Bottom half is more of a twist—while top half is more of a tilt boundary.) (Photo: M. Malone)

mixed character of edge- and screw components. The geometrical configuration of the partial twist boundary is generated as a projection of the pure twist boundary onto the inclined boundary plane, i.e. it is derived from the pure boundary by an affine transformation. It is not necessary that the boundary be planar. It may be a curved surface over which the dislocations continuously change their character. An example is given in Fig. 11.4/4.

d) Special Tilt Boundary

A tilt boundary is formed if the boundary surface is parallel to the axis of rotation, i.e. parallel to the lines formed by all the MS-points. Along these MS-lines the two crystals are in perfect register. Hence, a tilt boundary will pass, as far as possible, through these lines. Again different situations are possible:

1. The tilt boundary may be planar and pass through dense rows of MS-lines. The intersections of the boundary plane with the cell walls are again the dislocations. In this case the tilt boundary is formed by a series of parallel dislocation lines. The dislocations are of pure edge character and the boundary plane is perpendicular to the Burgers vectors (see Fig. 12.3/4).

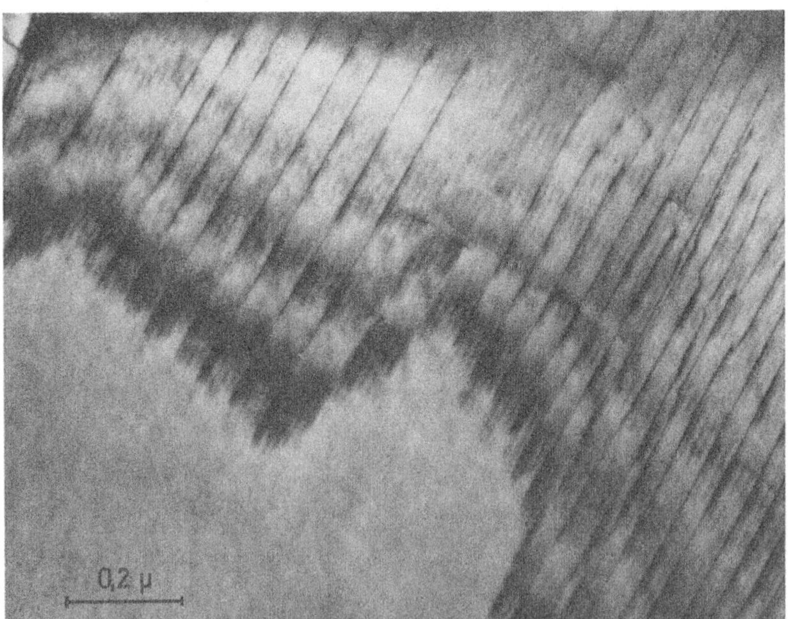

Fig. 11.4/5. Example of a stepped tilt boundary in iron. (Photo: A. Mascanzoni)

It was shown in Section 7.3 that two parallel edge dislocations could form a stable array, if glide were permitted but climb excluded. We now have the justification for this in that such arrangements are the origin of a stable tilt boundary.

2. The orientation of the tilt boundary may be random with respect to the cell structure. However, the boundary will not cut through the cell walls in an arbitrary manner, but will fit into the MS-lines as a stepped surface. Hence, such a boundary will be stepped and will consist of different types of edge dislocation (see Fig. 12.3/4). It does not have to be even approximately planar but can be curved and broken along the whole length (Fig. 11.4/5).

11.5. Foreign Dislocations

The "special" twist boundary is characterized by the fact that all the Burgers vectors lie in the same plane. Now we consider the case of one dislocation in the network having a Burgers vector which points in a direction out of that plane. We term such a dislocation a *foreign dislocation*. If the *b*-net is constructed following the methods given in Section 9.2, we see that the *foreign dislocation* produces a *step in the b-net* (Fig. 11.5/1). Stability considerations of the type given in Fig. 9.3/5

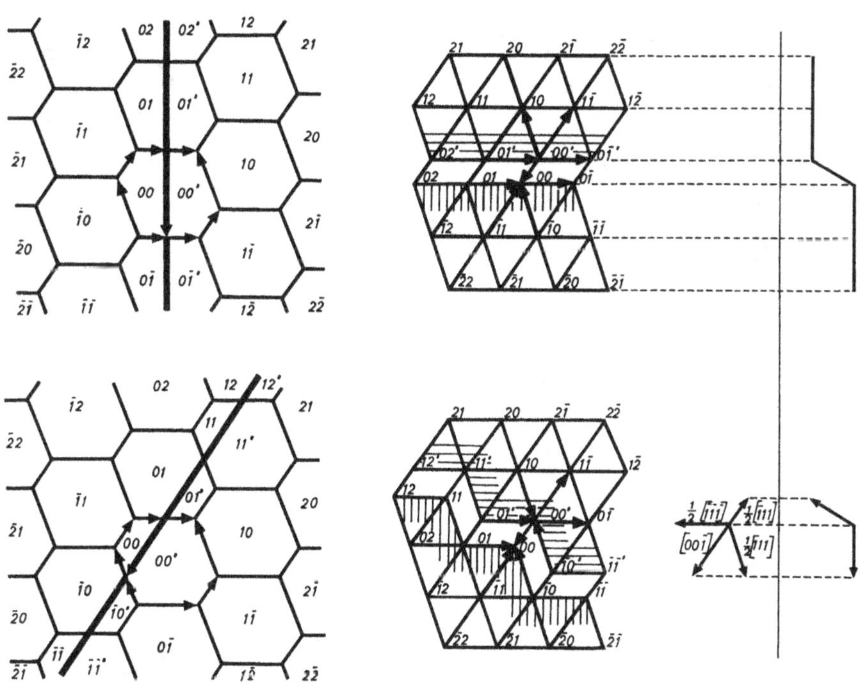

Fig. 11.5/1. Step in the *b*-net produced by a foreign dislocation

can be applied to such steps. Overhanging steps should be transformed into smoothly inclined ones.

11.6. "General" Subgrain Boundaries

In the case of the "general" subgrain boundary, the axis of rotation is crystallographically oriented in an arbitrary manner. In Section 11.2 it was shown that the *b*-net should form a plane perpendicular to the axis of rotation. However, this will be no longer possible with this axis in a general orientation, since the Burgers vectors can only be oriented in certain discrete directions. The planar *b*-net thus has to be approximated by a stepped one. In addition to a minimum of two linearly independant Burgers vectors in the "special boundary", two more are needed for the formation of two independant series of steps in the *b*-net. Thus, a really general boundary contains at least four independant series of dislocations. Naturally, boundaries with three independant series can also exist.

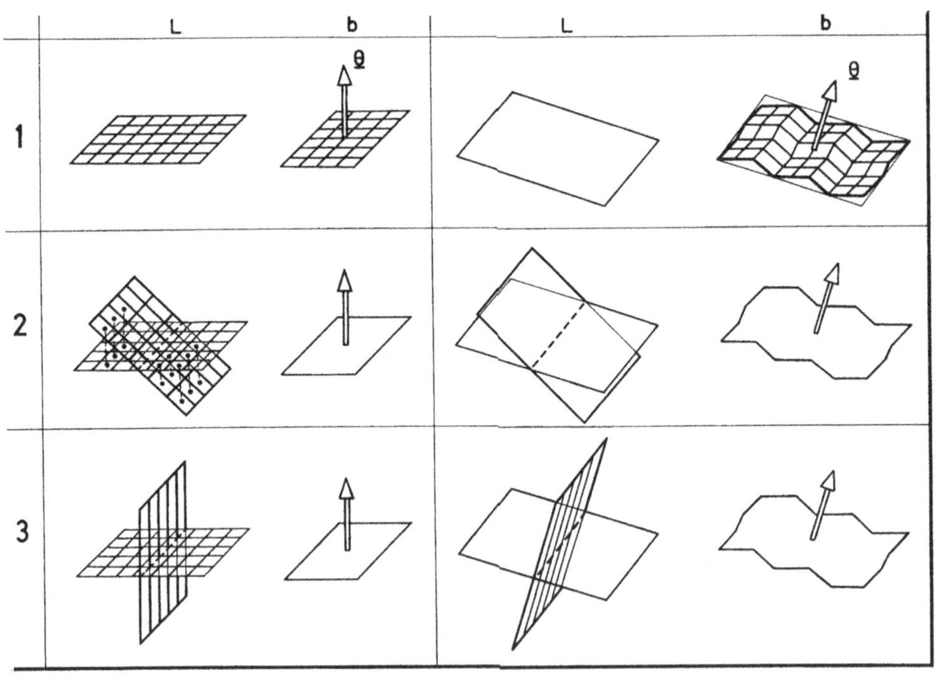

a b

Fig. 11.6/1 a and b. Summary of the different types of subgrain boundaries: a) special—and b) general boundaries. *1.* Pure twist —, *2.* partial twist —, *3.* tilt boundaries

It was shown in Eq. (11.2-4) that Frank's formula cannot be satisfied for b-points with $b_3 \neq 0$, i.e. such points do not have an x-line as an image. In our procedure involving the stepped b-net we disregard this condition. In Section 12.11 the justification will be given. (It will be shown that, in addition to the rotation $\boldsymbol{\theta}$, a translation of lattice 2 with respect to lattice 1 by an amount b_3 will be required.)

As far as the orientation of the boundaries (pure-, partial-, etc.) is concerned, the general- are similar to the special subgrain boundaries. However, in the general boundary not all the Burgers vectors are perpendicular to the rotation axis. Fig. 11.6/1 gives a summary of the various relative orientations.

11.7. The Energy of a Subgrain Boundary

The energy of a subgrain boundary is calculated for the case of a regular tilt boundary consisting of a row of similar parallel edge dislocations. In this manner we essentially follow Read (1953). The absolute value of the Burgers vectors shall be b, and h shall be the spacing between the dislocation lines. The angle of tilt θ then becomes:

$$\theta = \frac{b}{h}. \tag{11.7-1}$$

We consider a unit (e.g. 1 cm²) of the boundary surface. According to Section 6.4, the energy per unit length of an edge dislocation is:

$$\begin{aligned} E/\mathrm{cm} = \Gamma &= \frac{G b^2}{4\pi(1-\nu)} \ln\left(\frac{r_0}{b}\right) + \text{const.} \\ &= \Gamma_1 + \Gamma_2 \end{aligned} \tag{11.7-2}$$

Γ_1 = elastic energy of the dislocation in its surroundings,

Γ_2 = core energy.

The number of dislocations per unit surface is:

$$N = \frac{1\,(\mathrm{cm})}{h\,(\mathrm{cm})}. \tag{11.7-3}$$

Within a distance from the boundary which is much larger than h, the strain fields of the dislocations will cancel out and the crystals will be essentially undisturbed. Therefore, we choose $r_0 = h$ and obtain for the *elastic* energy of the boundary:

$$\begin{aligned} N \cdot \Gamma_1 &= \left(\frac{b}{h}\right) \cdot \frac{G b}{4\pi(1-\nu)} \ln\left(\frac{h}{b}\right) \\ &= E_0\, \theta \ln(1/\theta) \\ &= -E_0\, \theta \ln \theta . \end{aligned} \tag{11.7-4}$$

We assume the core energy per unit length Γ_2 to be proportional to b^2, i.e. proportional to the section of the cylinder which was cut out as "bad" material. On enclosing one factor b into θ and the other into E_0 we may write:

$$N\Gamma_2 = E_0 \cdot \theta \cdot A\,, \tag{11.7-5}$$

where A is a constant. Thus, the dependence of the boundary energy on the angle of tilt becomes:

$$\underline{E = E_0\,\theta\,(A - \ln\theta)\,.} \tag{11.7-6}$$

The plot of this function starts vertically ($dE/d\theta = \infty$ for $\theta = 0$) and attains a maximum value for:

$$\theta_{\max} = e^{(A-1)} \tag{11.7-7}$$

with a maximum value of the energy:

$$E_{\max} = E(\theta_{\max}) = E_0 \cdot e^{(A-1)}\,. \tag{11.7-8}$$

The maximum is attained for values of θ between 15 and 25°. The Eq. (11.7-6) is valied for all kinds of subgrain boundaries (not only tilt boundaries), if the constants E_0 and A are modified correspondingly. The above behaviour of the energy has been experimentally confirmed (Read, 1953, p. 192). Typical values are:

$$E_0 \approx 1{,}200 \text{ ergs/cm}^2, \quad E_{\max} \approx 600 \text{ ergs/cm}^2, \quad \theta_{\max} \approx 25°\,.$$

Referred to the maximum values, Eq. (11.7-6) becomes:

$$\frac{E}{E_{\max}} = \frac{\theta}{\theta_{\max}}\left[1 - \ln\frac{\theta}{\theta_{\max}}\right]. \tag{11.7-9}$$

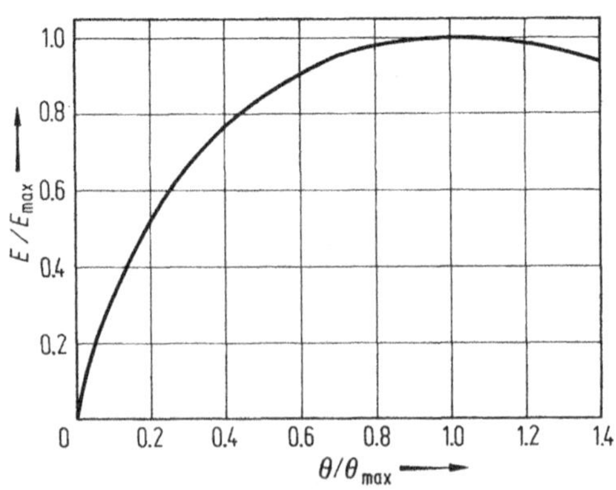

Fig. 11.7/1. Subgrain boundary energy [Eq. (11.7-9)]

This equation is no longer valid at large angles, θ, where the spacing between the dislocation cores becomes so small that the boundary becomes a high angle boundary.

11.8. Some Considerations of Recrystallization

On highly deforming a metal single crystal (e.g. by rolling), the hardness increases. On annealing for a fixed time at given temperatures, the hardness decreases as the annealing temperature increases. In the example (Fig. 11.8/1) of pure nickel, three different stages can be distinguished. The first ranging from room temperature to 250 °C is the recovery stage, the second from 250 to 400 °C is that of the "primary recrystallization" and at temperatures >400 °C there is the stage of "secondary recrystallization".

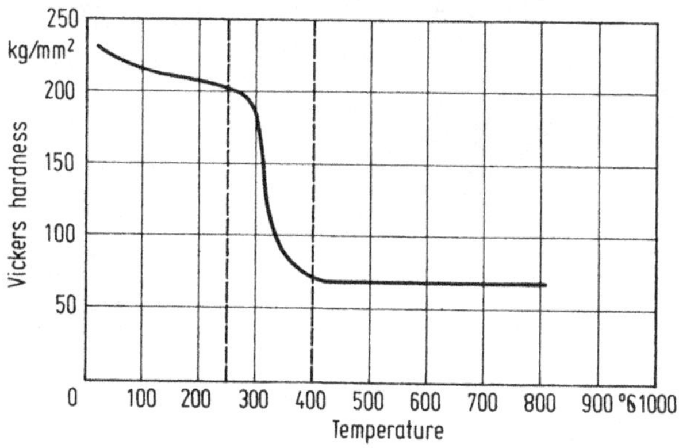

Fig. 11.8/1. Hardness of pure nickel during recrystallization

Electron microscopic examination of the internal state of the material reveals at room temperature a high dislocation density which, however, is not homogeneously distributed (Fig. 11.8/2). Small dislocation free areas are surrounded by a mass of highly entangled dislocations. During the recovery, the material becomes increasingly more clearly subdivided into cells ($\varnothing \approx 1\ \mu$) where the cell walls are formed by more or less regular dislocation networks. This situation is termed "polygonization" and hence, the recovery stage is due to polygonization.

At 250° distinct crystal grains appear at various spots which grow out into their surroundings (Fig. 11.8/3), and which are at least partially separated from their surroundings by discrete high angle boundaries. During the growth of these grains the dislocations are annihilated at the boundary (Fig. 11.8/4). At 400 °C all the grains have grown together, and

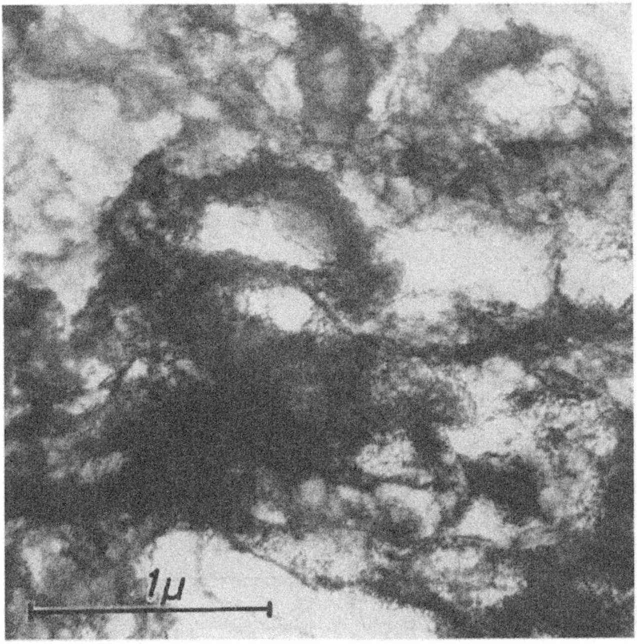

Fig. 11.8/2. Internal state of nickel after 80% cold rolling

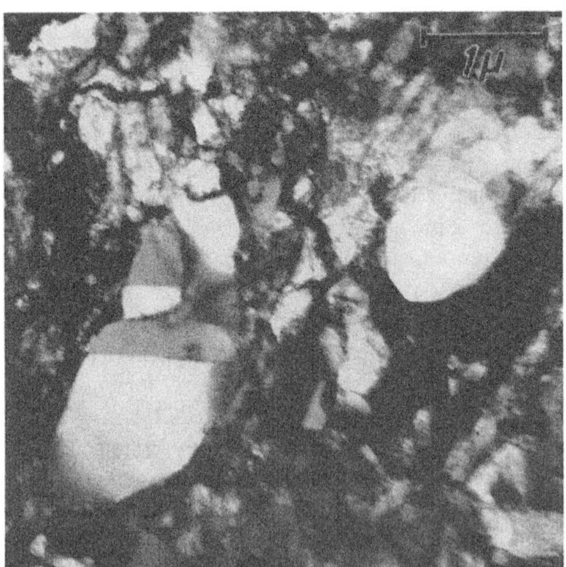

Fig. 11.8/3. Formation of nuclei of recrystallization in nickel at 250 °C

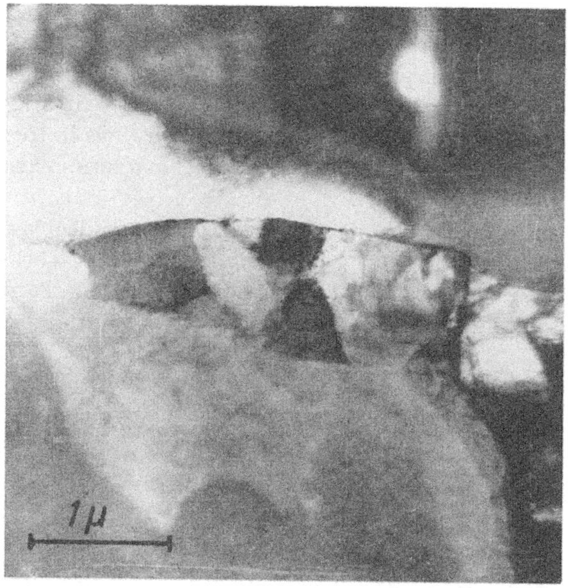

Fig. 11.8/4. Final stage of primary recrystallization in nickel. The last polygonized areas disappear

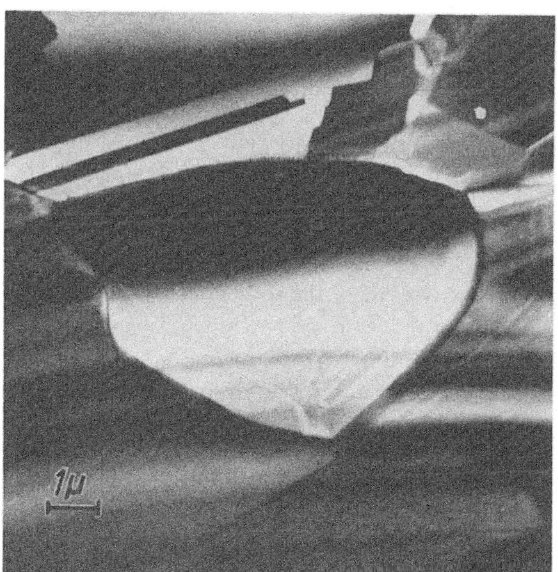

Fig. 11.8/5. State after the end of the primary recrystallization

nothing is left of the polygonized structure (Fig. 11.8/5). Hence, the material has lowered its hardness as well as its energy by eliminating the dislocations. At this point, the primary recrystallization stage comes to an end. In Fig. 11.8/5 lines marking the different stages of grain growth can be seen (analogous to annual growth rings in trees) and they are probably due to impurity segregation to boundaries existing at that time.

During further rearrangement, the material can decrease its energy only by decreasing the total grain boundary surface. This takes place during the stage of secondary recrystallization when the small grains disappear, the large grains grow and curved boundaries straighten out (Fig. 11.8/6).

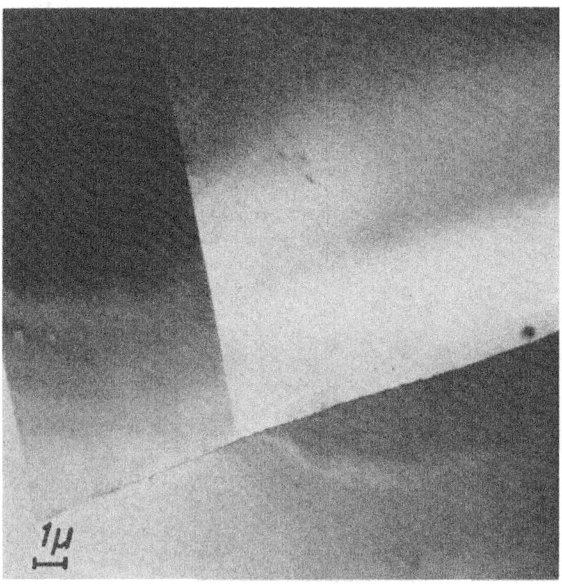

Fig. 11.8/6. Result of the secondary recrystallization

The question concerning the *nucleation of the crystal grains* now arises. Let us consider the energy content of a crystal grain in highly deformed surroundings. This energy consists of two contributions: the surface energy and the volume energy, i.e. the energy of the dislocation tangles. The value of the first is increasing and that of the second is decreasing while the grain grows. We idealize the grain as a sphere. In this case we obtain:

$$E = E_S^0 \, 4\pi \, r^2 - E_V^0 \, \frac{4\pi}{3} \, r^3 \qquad (11.8\text{-}1)$$

with E_S^0 = specific surface energy, i.e. energy per unit surface,
E_V^0 = specific volume energy.

According to Eq. (11.8-1) the energy of a crystal grain would first increase with increasing radius, attain a maximum at the critical radius r_c, and decrease afterwards. The critical radius r_c follows from $dE/dr = 0$ as:

$$r_c = \frac{2E_S^0}{E_V^0}. \tag{11.8-2}$$

One could imagine that, as in other nucleation problems, the grain would first have to be *forced* to grow while increasing its energy, in order to overcome the critical size; from this point on it could grow on its own. Let us remember, however, that the specific energy of a subgrain boundary depends on the misfit between the two sides of the boundary [Eq. (11.7-6)]. Thus, with a small misfit E_S^0, the critical radius r_c can be small. Here we assume E_V^0 to be essentially constant. As the dislocation density is very inhomogeneous in the deformed material it may happen that at certain spots the local radius of curvature of a dislocation free area is larger than r_c so that the nucleus can grow on its own. With the growth of the nucleus, the misorientation at the boundary, and hence E_S^0 and r_c, may increase (Fig. 11.8/7). Nevertheless, the radius of curva-

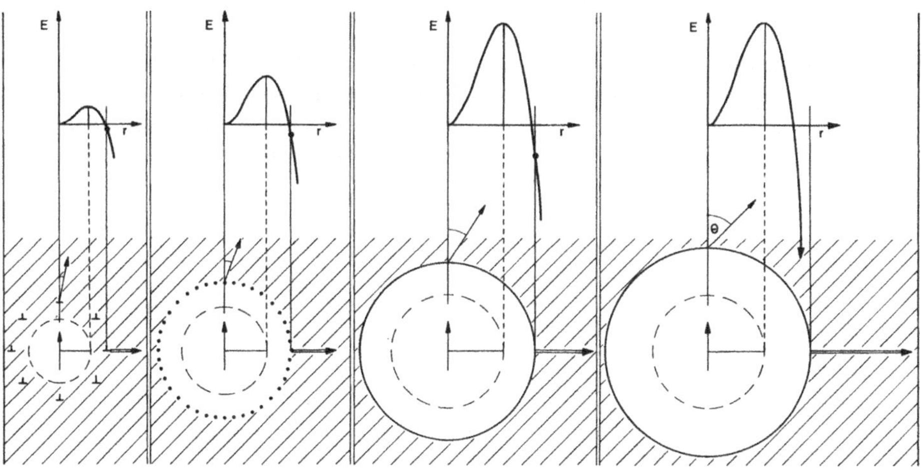

Fig. 11.8/7. Schematic representation of the growth of nuclei

ture of the boundary may still be larger than r_c whereby the grain continues to grow. At an angle of misfit of around 20°, the specific surface energy E_S^0 attains its maximum value. From this point on the critical radius remains essentially constant and the grain grows in an unrestrained manner until it meets another grain.

In the case where originally polycrystalline material was deformed, old high angle boundaries can act as nuclei for new grains. Here the driving force is given by the difference in volume energy (dislocation density) between the two sides of the old boundary. Once the boundary moves, the volume swept out is dislocation free, i.e. only one side of the boundary contains dislocations, and this sustains the driving force. As the boundary is relatively straight from the beginning, no critical radius has to be overcome (Bailey, 1960).

11.9. Limiting Cases

In Sections 11.2 to 11.6 it was shown how the theory of subgrain boundaries can be developed by employing Frank's formula together with the dualistic representation. Now we consider the limitations of this theory.

It is known that the accuracy of Frank's formula diminishes with increasing misfit between the two crystals. In addition, the formula is only valid if both crystals are rotated with respect to each other, i.e. for grain boundaries (with small misfit) but not for phase boundaries. Chapter 12 treats a generalization for arbitrary crystalline interfaces and Frank's formula proves to be a special case of a more general equation.

A further problem arises if a dislocation branches out from a network into one of the undisturbed crystals. On attempting to draw the b-net for this case, we find that there is no longer a "one-to-one" relationship between a b-net and an L-net (Fig. 11.9/1). The configuration of the b-net

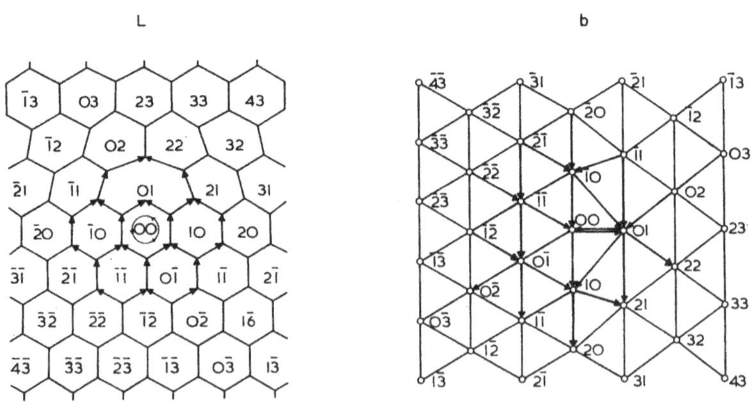

Fig. 11.9/1. Breaking away of a dislocation from a dislocation network

itself contains a "dislocation". The b-net is no longer topologically planar (i.e. it can no longer be spread into a plane). It is also not possible to define L-fields at the spot where the dislocation branches out. This difficulty will be cleared up in Section 14.3.

11.10. Dualistic Representation in Three Dimensions

On trying to join several subgrain boundaries, it can be seen that it is no longer possible to represent such an arrangement by a closed b-net. As every L-net has its own mesh width, i.e. its own $\boldsymbol{\theta}$, the different dislocation networks become interconnected by dislocations branching out of one boundary and being incorporated into the other. Thus, a common b-net would contain many "dislocations" (Section 11.9).

However, another possibility to extend the duality to three dimensions becomes apparent if, instead of a dislocation line, the whole surface of the subgrain boundary, and instead of the Burgers vector the $\boldsymbol{\theta}$-vectors are considered. Now, in place of the two-dimensional fields limited by dislocations, the three-dimensional subgrains are numbered. The b-space is replaced by a space of $\boldsymbol{\theta}$-vectors where the composition of

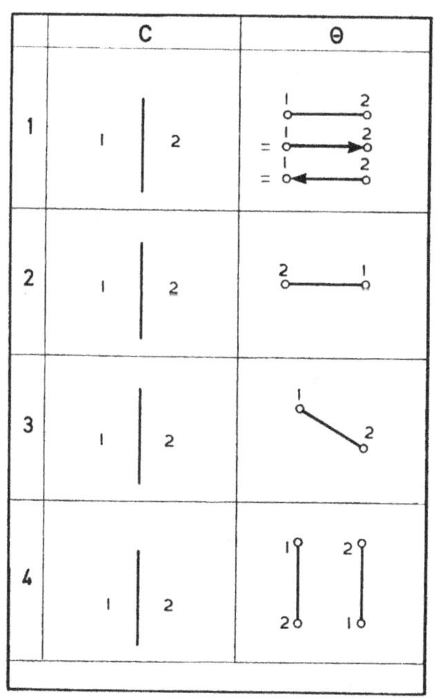

Fig. 11.10/1. Three-dimensional duality between subgrains and rotation vectors θ. *1.* Right-handed twist boundary, *2.* left-handed twist boundary, *3.* partial twist boundary, *4.* tilt boundary

these vectors forms a three-dimensional framework. Hence, instead of the b-nodes the nodes in this framework shall be numbered. The convention shall be fixed such that, at a junction of two θ-nodes (numbered N_1 and N_2), the node designated by N_1 (arbitrarily chosen) shall be the tail and the N_2-node the tip of the $\boldsymbol{\theta}$-vector whereby the $\boldsymbol{\theta}$-vector follows the normal right-handed screw convention. The subgrain N_1 shall be kept fixed and the subgrain N_2 shall be rotated by an amount $\boldsymbol{\theta}$. It can be seen that the relative orientation of the two subgrains is invariant while N_1 and N_2 are interchanged.

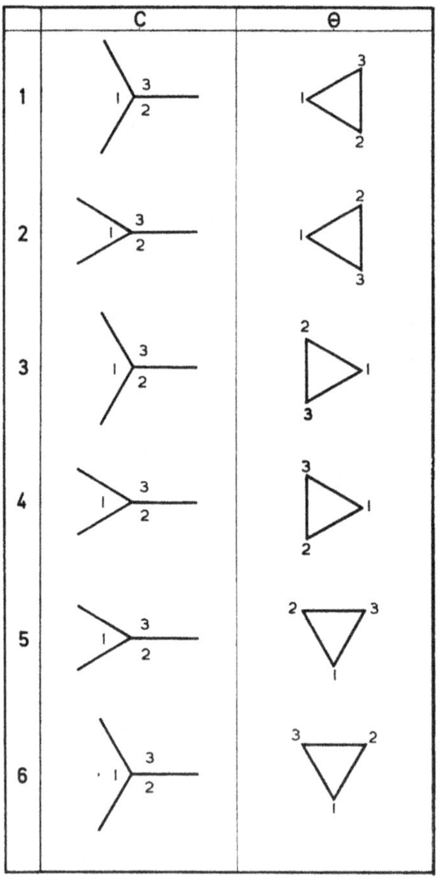

Fig. 11.10/2. Dualistic representation of the junction of three subgrains (Abbreviations: *Tw* twist boundary, *Tt* tilt boundary, *r* right-handed rotation, *l* left handed rotation). *1.* All three boundaries Tw r; *2.* (1, 2) Tt, (2, 3) Tw l, (3, 1) Tt; *3.* All three boundaries Tw l; *4.* (1, 2) Tt, (2, 3) Tw r, (3, 1) Tt; *5.* (1, 2) Tw l, (2, 3) Tt, (3, 1) Tw r; *6.* All three boundaries Tt

Fig. 11.10/1 represents different kinds of subgrain boundaries. The boundary surface is oriented normal to the plane of the paper.

The following dualities exist between various entities in the C-space (crystal space) and in the θ-space:

C-space		θ-space
1. A subgrain (3 dimensions, abbreviated 3D)	corresponds to	a point where several $\boldsymbol{\theta}$-vectors meet (0D)
2. A subgrain boundary (2D)	corresponds to	a $\boldsymbol{\theta}$-vector (1D)
3. An intersection of subgrain boundaries in a line (1D)	corresponds to	a polygon of $\boldsymbol{\theta}$-vectors (2D)
4. An intersection of several subgrain boundaries in one point (0D)	corresponds to	a polyhedron of $\boldsymbol{\theta}$-vectors (3D)

Examples of possible subgrain boundaries meeting in a line are given in Fig. 11.10/2.

This kind of duality between subgrains and $\boldsymbol{\theta}$-vectors is an approximation of the same type as is represented by Frank's formula. Rotations can never be added exactly as vectors. In reality they are noncommutative tensors. The smaller the angles $\boldsymbol{\theta}$, the better, however, becomes the approximation.

Chapter 11. Summary and Discussion

In this chapter, the theory of subgrain boundaries has been treated, i.e. the boundaries between two crystals of the same kind slightly rotated with respect to one another. These boundaries are formed of dislocation networks. Their description is based on the interpretation of Frank's formula [Eq. (11.2-1)] which represents an algebraic relation between the b-space and the L-space. Frank's formula can be interpreted as an imaging relation between a plane in the b-space and the complete crystal-space (L-space). The b-plane passes through the origin of the b-space and is perpendicular to the axis of rotation $\boldsymbol{\theta}$. For every point in the b-plane, there exists a corresponding x-line parallel to $\boldsymbol{\theta}$ in the crystal space.

If Frank's formula is quantized [Eq. (11.2-5)], the discrete b-points, which are possible combinations of Burgers vectors, are imaged as a set of discrete x-lines parallel to $\boldsymbol{\theta}$. These lines are composed of minimum strain points, abbreviated to MS-points. The separation of the MS-lines by cell walls produces a cell structure. The possible dislocation networks forming the boundary can be understood as sections through this cell structure. A classification of subgrain boundaries is developed on this basis (Section 11.3). For the special orientations of $\boldsymbol{\theta}$, the situation is uniquely described by the quantized Frank's formula, limited by the fact, however, that Frank's formula itself is an approximation. The procedure also enables the study of dislocation networks containing partial dislocations (Section 11.4).

For the general orientations of $\boldsymbol{\theta}$, the b-net becomes stepped, as "foreign" dislocations (Section 11.5) are introduced into the boundary. Frank's formula is violated by this procedure. Nevertheless, the duality relations do hold for the correlations between Burgers vectors and dislocation lines. The (b^1, b^2)-coordinate axes of the b-net are perpendicular to the (x^1, x^2)-axes of the L-net, since $\boldsymbol{b}$ stays perpendicular to $\boldsymbol{x}$.

It was shown in the Summary of Chapters 7–9 how the topological properties of the b-net could be derived from the configuration of the L-net. The justification for such a procedure lies in the fact that every dislocation arrangement of some complexity will tend to approximate to a subgrain boundary.

The above theory of subgrain boundaries is not valid for large angles of rotation θ, i.e. in the domain of general grain- and phase boundaries. The theory will be extended into this domain in the following chapters. On the other hand, the duality relations cannot be fully applied if the L-net is no longer topologically planar (branching of a dislocation out of a network, Section 11.9). This problem will also be solved in Section 14.3. In Section 11.10 it was shown that it is possible to generalize the duality relations between the individual subgrain boundary and its b-net to a dualistic representation of a polygonized structure.

Suggestions for Problems

Many cases of special and general subgrain boundaries can be constructed and discussed in detail. The configurations given in the article by Amelinckx and Dekeyser (1959) which are represented in Thompson's notation, can be transformed into the dualistic representation. With respect to the energy of subgrain boundaries, exercises can be found in the books of Read (1953) and of Weertman and Weertman (1964).

12. General Geometrical Theory of Crystalline Interfaces

12.1. Introduction

A boundary surface between any two crystals will take up such a position that the crystals exhibit more or less optimum matching. Our task will be to formulate the notion of the "matching" of two periodic structures in a correct mathematical manner. A start has already been made in the case of the MS-points of the quantized version of Frank's formula [Eq. (11.2-5)].

We idealize the two crystals as two interpenetrating mathematical translation lattices, i.e. we disregard space filling atoms. Then we look for positions of "best fit" in these interpenetrating lattices. It will be shown in the next sections how this is achieved. When such positions have been distinguished, a prospective boundary can be placed through them. A special example was the "honeycomb"-structure of the sub-grain boundaries (Section 11.2). For monatomic materials (e.g. fcc or bcc metals) the interpenetrating lattices are understood as being *two* sets of positions for *one* set of atoms, which, depending on the set chosen, can form either crystal 1 or crystal 2. Once the boundary is fixed, the atoms may be fitted into set 1 on the one side, and into set 2 on the other. The boundary itself remains as an approximately two-dimensional remnant of the two interpenetrating three-dimensional lattices.

The interpenetrating lattices with the positions of "best fit" represent a complete description of all the *possible* boundaries between two crystals of given structures and orientations. As soon as the boundary is chosen and "realized", the remainder of the interpenetrating lattices lose their meaning.

12.2. The Coincidence-Site Lattice

Let us consider two interpenetrating translation lattices (lattice 1 and lattice 2). In our investigations, we usually assume lattice 1 to be fixed and all the changes such as: translation, rotation, etc. are performed with lattice 2. With the relative orientation of the two lattices given, lattice 2 is translated in such a way that one of its points coincides with a point in lattice 1. That point shall be termed a "lattice coincidence site". An atom located there is in an unstrained position in both lattices. For this reason we regard this point as a "point of best fit".

We may ask whether other examples of lattice coincidence occur within the interpenetrating lattices: In effect there exist either none, or one (the one which was arbitrarily produced) or, due to the periodicity of the two lattices, an infinite number of them. This infinity forms a lattice of coincidence sites, i.e. a *coincidence site lattice.*

Taking the example of two equal, two-dimensional square lattices, we examine the orientations (of lattice 2 if lattice 1 is fixed) for which coincidence site lattices exist, and we investigate the structures of these coincidence site lattices. We hereby fix the coordinate system such that the coordinates of all the lattice points (of lattice 1) are integers (crystal coordinate system Section 2.2). Initially, both lattices are placed into the same orientation and brought to full coincidence. Then, choosing a lattice point as origin, we rotate lattice 2 by the angle θ in the positive sense around the coordinate origin. In this way the pattern formed of lattice points of both lattices varies with θ. After a rotation of 90° both lattices once more coincide completely. However, already with rotations of up to 45°, we obtain all the possible patterns if we regard those patterns as being identical which can be superposed by rotation. Thus, the patterns for θ and for $90° - \theta$ are identical.

Inside lattice 1 we delimit a sector of 45° by the two mirror axes [1, 0] and [1, 1]. Then, we subdivide this sector into two halves of 22.5° (Fig. 12.2/1). One of these sectors is limited by the [1, 0]-axis and the

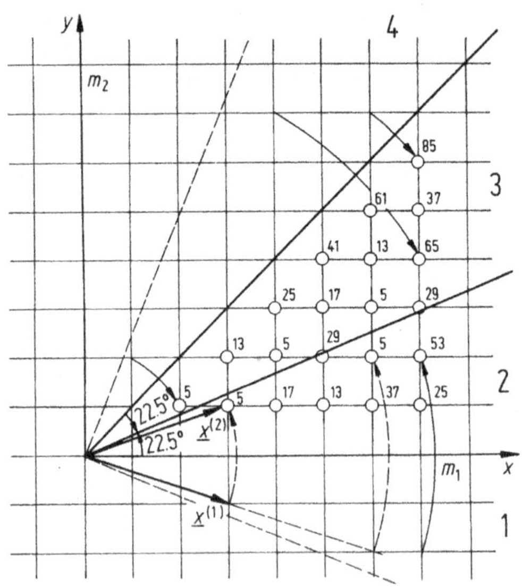

Fig. 12.2/1. Determination of the Σ-values for rotation on the {100}-plane in the cubic structure (see Table 13.3/T1, No. 1–22)

other by the [1, 1]-axis. Any point from outside such a sector can be transferred into a point inside by a rotation of $\theta < 45°$. The angle between the mirror axis and the position vector of one of the two points related in this way is $\theta/2$. Fig. 12.2/1 shows that the structure of the pattern does not depend on whether a point of sector 1 is rotated into one of sector 2 by $+\theta$ or vice versa. The orientation of the pattern, however, is not the same. On rotation by positive angles θ the corresponding lattice coincidence sites lie in sectors 2 and 4, and for negative angles θ in sectors 1 and 3.

In our example of the square lattices the coincidence site lattice must be itself a square lattice, since for every point rotated over the x-axis a similar point must be rotated over the y-axis. A measure of the coincidence site lattice is given by the surface of its unit cell. When considering whether a point within a sector (e.g. sector 2) is a possible lattice coincidence site close to the origin, it must be assured that its coordinates (x, y) do not contain common factors, since this would mean that a fraction of the position vector would again lead to a coincidence site. In particular, the two coordinates should not be even.

For a chosen point (x, y) of lattice 1, for example (Fig. 12.2/2), the surface of the square of the position vector is given by:

$$\Sigma' = x^2 + y^2. \tag{12.2-1}$$

For $(x, y) = (3, 1)$ $\Sigma' = 10$.

Both coordinates 3 and 1 are odd numbers, hence the sum of their squares is an even number. It is evident, however, that the square (formed by the origin and three coincidence sites) contains a coincidence site in its centre. $[(3/2, 1/2) + (-1/2, 3/2) = (1, 2)]$. For this central point $\Sigma' = (1)^2 + (2)^2 = 5$ and this is the smallest unit for the given angle of rotation

$$\theta = 2 \arctan (y/x) \tag{12.2-2}$$

($= 36°\ 52.2'$ in our example).

If Σ' is even, then its value has to be divided by 2 in order to obtain the smallest unit.

$$\Sigma = \frac{1}{\alpha} \Sigma', \tag{12.2-3}$$

with $\alpha = 2$ if Σ' is even, and $\alpha = 1$ for Σ' odd. Σ is called the "generating function".

For the point (4, 1), $\Sigma' = 17$ and the centre of the square $[(2, 1/2) + (-1/2, 2) = (3/2, 5/2)]$ is not a lattice coincidence site, as its coordinates are not integers. Care must be taken with large coordinates since, occasionally, certain lattice coincidence sites can belong to several coincidence site lattices.

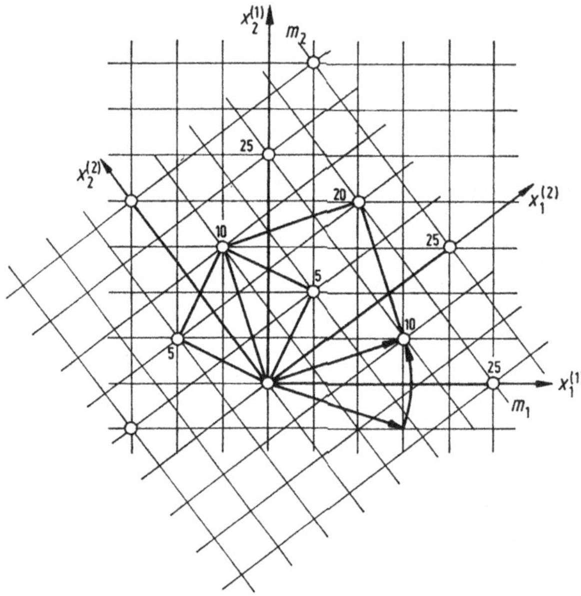

Fig. 12.2/2. Special coincidence site lattice [$\Sigma = 5$, $\theta = 36°\ 52.2'$, $\tan(\theta/2) = 1/3$] (Table 13.3/T 1, No. 19)

This example of a coincidence site lattice has been treated here in such detail because, in Section 13.3, it will become important in a more general form for the study of the periodic or nonperiodic structure of patterns.

The coincidence site lattice is not yet the full solution of the problem of our search for locations of best fit. It still contains serious deficiencies.

First of all, the knowledge of the position of a coincidence site tells us nothing about its surroundings except that the pattern of lattice points must be prieodic (with the periodicity of the coincidence site lattice). The converse is not true, since periodic patterns may exist without coincidence sites.

Secondly, as Fig. 12.2/3 shows, it is practically impossible to identify a coincidence site in a more complicated pattern. The reason lies in the fact that the concept of the coincidence site contains no measure of the spacing between two points. Whether two points coincide or are 0.0001 Å apart may be physically insignificant. Mathematically, however, the situations are fundamentally different.

Thirdly, let us assume the orientation of lattice 2 to be such that there exists quite a dense coincidence site lattice. This lattice is completely destroyed by an infinitesimal change in the orientation of

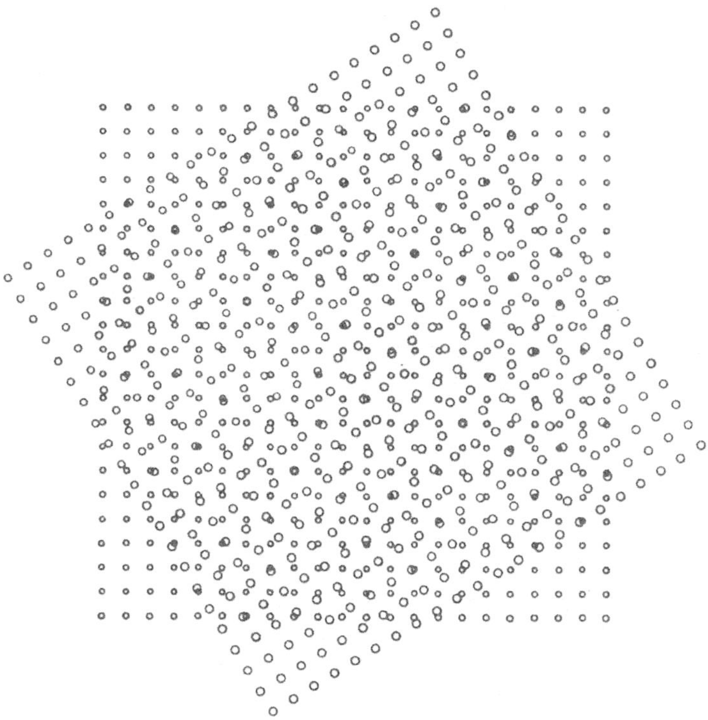

Fig. 12.2/3. Superposition of two square lattices. The picture contains four coincidence sites

lattice 2, although the change in the physical situation is expected to be infinitesimal. In addition, there are an infinity of orientations for which no coincidence site lattice exists, namely those for which, in the case of the square lattices, $\tan(\theta/2)$ is irrational. Hence, the coincidence site lattice is extremely discontinuous in its behaviour for changes of the orientation of lattice 2. In contrast to this, the MS-points (Section 11.2) are continuous functions of θ.

In order to be able later to generalize the concept of the coincidence site lattice we have to express the facts somewhat more mathematically. Each of the two lattices can be understood as a translation group constructed by means of three linearly independant basic vectors. We may correlate the two lattices point by point by a linear homogeneous transformation $\boldsymbol{A}$

$$\boldsymbol{x}^{(2L)} = \boldsymbol{A}\boldsymbol{x}^{(1L)} \qquad |\boldsymbol{A}| \neq 0. \tag{12.2-4}$$

The index L means "lattice point". In our example, $\boldsymbol{A}$ is a rotation, but it may be a general affine transformation (expansion, rotation, shear, etc.). The determinant $|\boldsymbol{A}| \neq 0$ means that both lattices shall be defined, with respect to the number of dimensions, in the same space, i.e., for example, a three-dimensional lattice shall not be imaged onto a two-dimensional one.

The whole set of the lattice points may be called an "equivalence class" or simply a "class". Hence, the two lattices are two related classes of points [related by Eq. (12.2-4)]. Every lattice point is therefore an "element" of one of the two classes. *Lattice coincidence sites are thus coincidences of elements of related classes.* [They are not coincidences of related elements, since by the relation in Eq. (12.2-4), a coincidence is only possible at the origin.]

In the next section we shall generalize the notion of the coincidence site lattice based on the above reflections.

12.3. The 0-Lattice

In Section 12.2 the lattice coincidence sites were defined as being coincidences of elements of related point classes, where the relation was given by Eq. (12.2-4).

The whole space of a crystal does not only consist of lattice points but also of the interstices between them; or expressed mathematically, not only of the translation group but also of all their cosets. A coset is formed of a position vector of a point within a unit cell plus the complete translation group.

When we choose the crystal coordinate system as being given by the unit vectors of lattice 1, then all the lattice points are indicated by integral coordinates. Considering the (cyrstal) coordinates of an arbitrary point e.g. (12.138, 7.243, −4.421) we designate as *external coordinates* the integers (12, 7, −5) and as *internal coordinates* (0.138, 0.243, 0.579). Hence, the external coordinates designate the unit cell and the internal coordinates the positions within that unit cell. The internal coordinates x_i shall always have values such that $0 \leqq x_i < 1$.

In the crystal coordinate system all the points of a coset have the same internal coordinates and so may be designated by them. Every coset is represented inside every unit cell (Fig. 12.3/1). Together with the translation group having internal coordinates (000), the infinite number of cosets forms a "partition": a subdivision of the whole space into "equivalence classes".

Through the equation

$$\boldsymbol{x}^{(2)} = \boldsymbol{A}\boldsymbol{x}^{(1)} \qquad |\boldsymbol{A}| \neq 0 \tag{12.3-1}$$

every point in crystal 1 is related to a point in crystal 2, and hence every equivalence class of crystal 1 is related to one in crystal 2. The related class in crystal 2 has the same internal coordinates in the coordinate system of crystal 2 as the original class in crystal 1. While Eq. (12.2-4) was only formulated for lattice points, Eq. (12.3-1) includes all the points in the space. As such, the two interpenetrating lattices represent two different partitions of the same space.

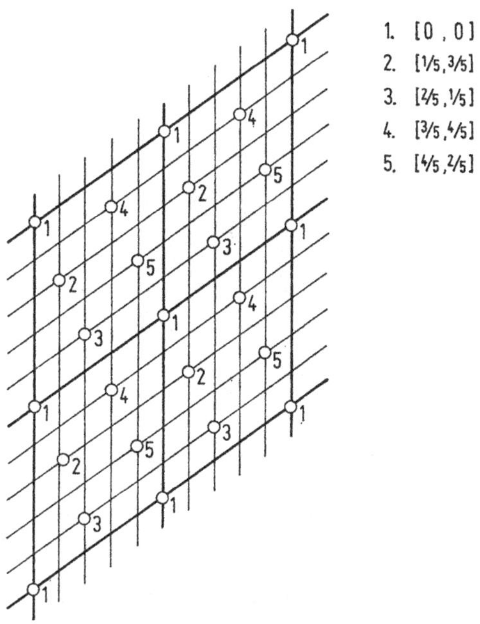

Fig. 12.3/1. Equivalence classes (Table 13.3/T 1, No. 43)

We now define "0-points", the generalized lattice coincidence sites, as *coincidences of elements of related equivalence classes*, no matter what may be the values of the internal coordinates of these classes.

Hence, the lattice coincidence sites are special 0-points with the internal coordinates (000). Roughly speaking, the 0-points can be understood as coincidences of internal coordinates or coincidences of points which are in equivalent positions in the two crystals.

We shall now derive an equation for the 0-points based on the above definition. We start from an arbitrary point $\boldsymbol{x}^{(1)}$ in crystal 1 with arbitrary external and internal coordinates. We term as $C^{(1)}$ the equivalence class of that point $\boldsymbol{x}^{(1)}$. The corresponding point $\boldsymbol{x}^{(2)}$ in the related class $C^{(2)}$ is given by

$$\boldsymbol{x}^{(2)}(C^{(2)}) = \boldsymbol{A}\big(\boldsymbol{x}^{(1)}(C^{(1)})\big). \qquad (12.3\text{-}2)$$

Alternatively, we can, starting from $\boldsymbol{x}^{(1)}(C^{(1)})$, find other elements $\boldsymbol{e}(C^{(1)})$ of the class $C^{(1)}$ by adding translation vectors $\boldsymbol{t}^{(1)}$ of lattice 1 to $\boldsymbol{x}^{(1)}$. Hence:

$$\boldsymbol{e}(C^{(1)}) = \boldsymbol{x}^{(1)}(C^{(1)}) + \boldsymbol{t}^{(1)}. \tag{12.3-3}$$

If the points defined by Eqs. (12.3-2) and (12.3-3) coincide, we designate that coincidence point by $\boldsymbol{x}^{(0)}$ (Fig. 12.3/2):

$$\boldsymbol{x}^{(2)}(C^{(2)}) = \boldsymbol{x}^{(1)}(C^{(1)}) + \boldsymbol{t}^{(1)} = \boldsymbol{x}^{(0)}(C), \tag{12.3-4}$$

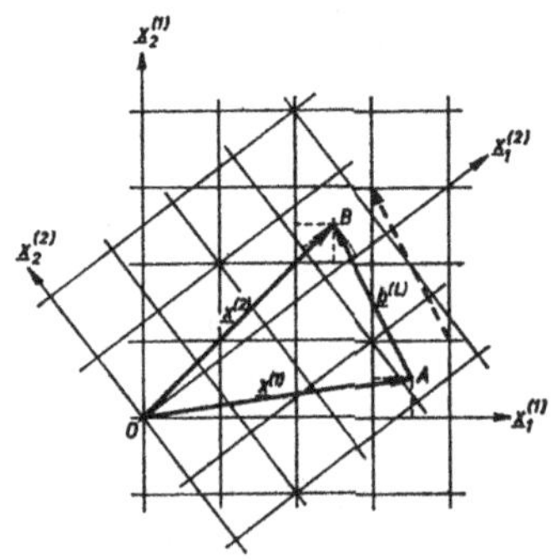

Fig. 12.3/2. Coincidence of elements of related equivalence classes

where C means that the point belongs simultaneously to both related classes $C^{(1)}$ and $C^{(2)}$. Together with Eq. (12.3-1) we obtain:

$$\boldsymbol{x}^{(0)} = \boldsymbol{A}^{-1}\,\boldsymbol{x}^{(0)} + \boldsymbol{t}^{(1)}$$

or

$$(\boldsymbol{I} - \boldsymbol{A}^{-1})\,\boldsymbol{x}^{(0)} = \boldsymbol{t}^{(1)} \tag{12.3-5}$$

(I = identity = unit transformation).

As we shall see later (Section 12.5) that all the possible translation vectors $\boldsymbol{t}^{(1)}$ of lattice 1 form the b-lattice, we already change the notation here; hence $\boldsymbol{t}^{(1)} := \boldsymbol{b}^{(L)}$ (L means lattice point). Thus:

$$\boxed{(\boldsymbol{I} - \boldsymbol{A}^{-1})\,\boldsymbol{x}^{(0)} = \boldsymbol{b}^{(L)}}\,. \tag{12.3-6}$$

All the 0-points are solutions of Eq. (12.3-6). We shall see that this equation is the basis of the whole geometrical theory of crystalline interfaces.

The general procedure for solving the equation is as follows:

1. Choice of the coordinate system.

2. Formulation of the transformation $\boldsymbol{A}$ which relates the two lattices.

3. Formulation of the basic Eq. (12.3-6) by matrix operations.

4. If the determinant $|\boldsymbol{I}-\boldsymbol{A}^{-1}| \neq 0$, the solution is

$$\boldsymbol{x}^{(0)} = (\boldsymbol{I}-\boldsymbol{A}^{-1})^{-1}\, \boldsymbol{b}^{(L)}. \tag{12.3-7}$$

The problems arising when $|\boldsymbol{I}-\boldsymbol{A}^{-1}| = 0$ will be discussed in Section 12.9.

In order to get acquainted with the procedure we calculate the example of Fig. 12.2/2, where the difference between the coincidence site lattice and the 0-lattice becomes evident. As the basic system we choose the orthogonal coordinate system of crystal 1. We formulate the problem two-dimensionally. The transformation $\boldsymbol{A}$ of Eq. (12.3-1) in this case is a rotation $\boldsymbol{R}$:

$$\boldsymbol{A} = \boldsymbol{R} = \begin{pmatrix} \cos\theta & -\sin\theta \\ \sin\theta & \cos\theta \end{pmatrix}. \tag{12.3-8}$$

From this follows:

$$\boldsymbol{A}^{-1} = \begin{pmatrix} \cos\theta & \sin\theta \\ -\sin\theta & \cos\theta \end{pmatrix} \tag{12.3-9}$$

and

$$(\boldsymbol{I}-\boldsymbol{A}^{-1}) = \begin{pmatrix} 1-\cos\theta & -\sin\theta \\ \sin\theta & 1-\cos\theta \end{pmatrix}. \tag{12.3-10}$$

If $|\boldsymbol{I}-\boldsymbol{A}^{-1}| \neq 0$ the solution is given by Eq. (12.3-7). In our example the determinant is:

$$|\boldsymbol{I}-\boldsymbol{A}^{-1}| = 2\,(1-\cos\theta) \tag{12.3-11}$$

and the inverse matrix $(\boldsymbol{I}-\boldsymbol{A}^{-1})^{-1}$ becomes:

$$(\boldsymbol{I}-\boldsymbol{A}^{-1})^{-1} = \begin{pmatrix} 1/2 & (1/2)\,\mathrm{cotan}\,(\theta/2) \\ -(1/2)\,\mathrm{cotan}\,(\theta/2) & 1/2 \end{pmatrix}. \tag{12.3-12}$$

As in the crystal coordinate system the two basic translation vectors are given by the coordinates (1, 0) and (0, 1), the unit vectors of the 0-lattice are the cloumn vectors of the matrix (12.3-12), i.e.

$$\boldsymbol{u}_1^{(0)} = \big(1/2,\ -(1/2)\,\mathrm{cotan}\,(\theta/2)\big) \quad \text{and} \quad \boldsymbol{u}_2^{(0)} = \big((1/2)\,\mathrm{cotan}\,(\theta/2),\ 1/2\big).$$

The geometrical interpretation of this result is given in Fig. 12.3/3. $\boldsymbol{u}_1^{(0)}$ is determined by the point of intersection of a straight line, which subtends an angle $\theta/2$ on the negative y-axis, with the vertical line at $x = 1/2$, and correspondingly $\boldsymbol{x}_2^{(0)}$ with the x-axis Fig. 12.3/4 shows two tilt boundaries selected from the same 0-lattice.

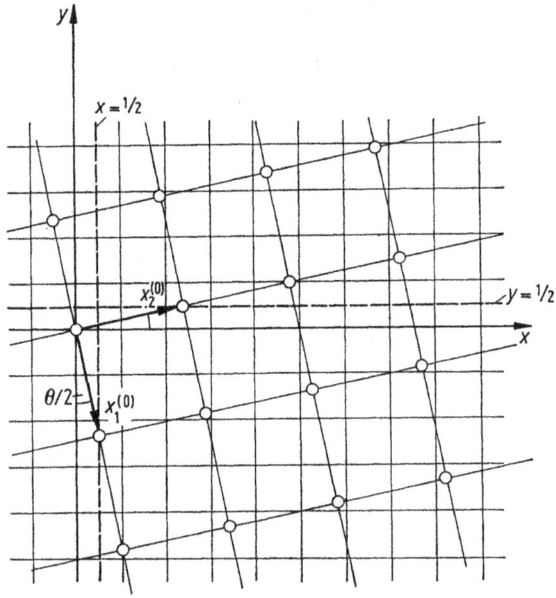

Fig. 12.3/3. Construction of the O-lattice according to Eq. (12.3-12)

In the special case of the coincidence site lattice Fig. 12.2/2, $\cot(\theta/2) = 3$ and

$$\begin{pmatrix} x_1^{(0)} \\ x_2^{(0)} \end{pmatrix} = \begin{pmatrix} 1/2 & 3/2 \\ -3/2 & 1/2 \end{pmatrix} \begin{pmatrix} b_1^{(L)} \\ b_2^{(L)} \end{pmatrix}. \tag{12.3-13}$$

The unit vectors of the O-lattice are $\boldsymbol{u}_1^{(0)}$ $(1/2, -3/2)$ and $\boldsymbol{u}_2^{(0)}$ $(3/2, 1/2)$ (see Fig. 12.5/1). Every second O-point is a lattice coincidence site. The other O-points belong to the equivalence class **[**1/2, 1/2**]*** (Fig. 12.5/1).

In the case of $\cot(\theta/2) = 4$ the following equivalence classes contain coinciding elements: **[**0, 0**]**, **[**1/2, 0**]**, **[**0, 1/2**]** and **[**1/2, 1/2**]**, which may be shown as an exercise.

We see that the O-lattice, in contrast to the general coincidence site lattice, is a continuous function of the transformation $\boldsymbol{A}$ (e.g. of θ) and that an O-lattice also exists where no coincidence site lattice is present. A slight change of $\boldsymbol{A}$ leads to the instantaneous disappearance of a coincidence site lattice, while the O-lattice undergoes a smooth change; a property which is common to the MS-points (Section 11.2). It will be shown in the next section that Eq. (12.3-6) can be understood as a generalization of the quantized version of Frank's formula [Eq. (11.2-5)], and that, hence, the MS-points may also be looked on as special O-points.

* The equivalence classes are indicated by boldface brackets **[**...**]**.

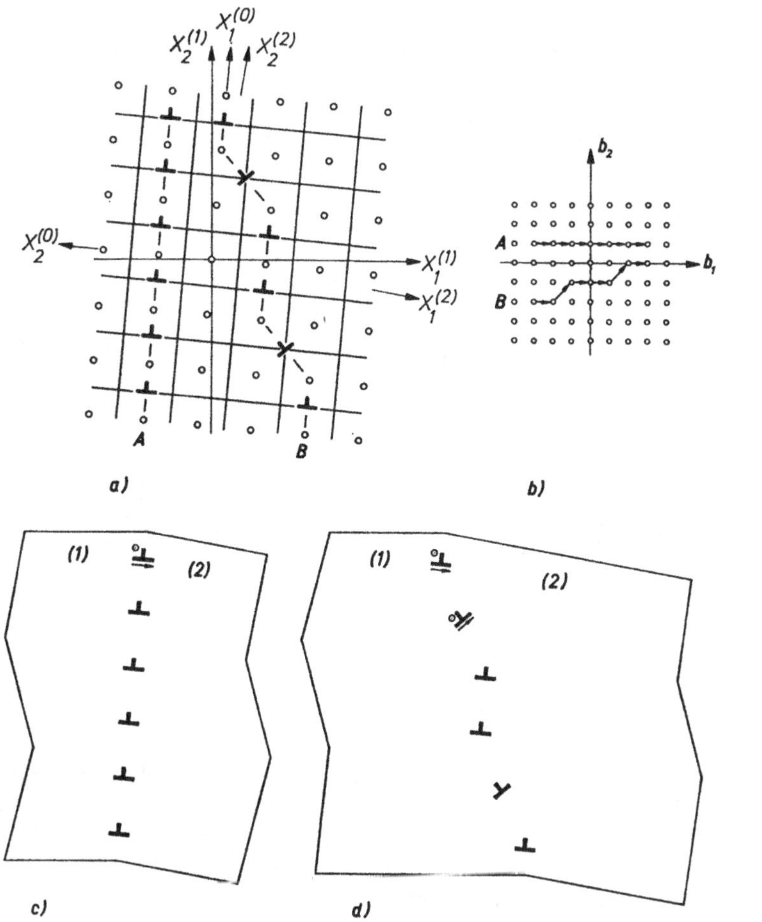

Fig. 12.3/4. Application of the 0-lattice to the construction of a symmetrical and an unsymmetrical tilt boundary

12.4. The 0-Lattice and Frank's Formula

In Section 11.2 the theory of subgrain boundaries was developed from Frank's formula. If we can show that Eq. (12.3-6) is a natural generalization of that formula, then we find that there exists a continuous transition from low-angle boundaries to general crystalline interfaces. In terms of coordinates Frank's formula is:

$$\begin{aligned} b_1 &= -\theta x_2 \\ b_2 &= \theta x_1 \\ b_3 &= 0 \end{aligned} \qquad (12.4\text{-}1)$$

where $\boldsymbol{b}(b_1\, b_2\, b_3)$ represents the discrete points of the b-net and $\boldsymbol{\theta}$ is placed in the z-axis. In matrix notation this equation becomes:

$$\begin{pmatrix} b_1 \\ b_2 \\ b_3 \end{pmatrix} = \begin{pmatrix} 0 & -\theta & 0 \\ \theta & 0 & 0 \\ 0 & 0 & 0 \end{pmatrix} \begin{pmatrix} x_1 \\ x_2 \\ x_3 \end{pmatrix}. \qquad (12.4\text{-}2)$$

On the other hand, according to the procedure given in Section 12.3, for a rotation around the z-axis, $\boldsymbol{A}$ is given by:

$$\boldsymbol{A} = \begin{pmatrix} \cos\theta & -\sin\theta & 0 \\ \sin\theta & \cos\theta & 0 \\ 0 & 0 & 1 \end{pmatrix} \qquad (12.4\text{-}3)$$

and Eq. (12.3-6)

$$\begin{pmatrix} b_1^{(L)} \\ b_2^{(L)} \\ b_3^{(L)} \end{pmatrix} = \begin{pmatrix} 1-\cos\theta & -\sin\theta & 0 \\ \sin\theta & 1-\cos\theta & 0 \\ 0 & 0 & 0 \end{pmatrix} \begin{pmatrix} x_1^{(0)} \\ x_2^{(0)} \\ x_3^{(0)} \end{pmatrix}. \qquad (12.4\text{-}4)$$

If θ is small, $\cos\theta \approx 1$ and $\sin\theta \approx \theta$, and Eq. (12.4-4) becomes:

$$\begin{pmatrix} b_1^{(L)} \\ b_2^{(L)} \\ b_3^{(L)} \end{pmatrix} = \begin{pmatrix} 0 & -\theta & 0 \\ \theta & 0 & 0 \\ 0 & 0 & 0 \end{pmatrix} \begin{pmatrix} x_1^{(0)} \\ x_2^{(0)} \\ x_3^{(0)} \end{pmatrix} \qquad (12.4\text{-}5)$$

which is identical with Eq. (12.4-1). The $\boldsymbol{b}^{(L)}$-vectors representing translation vectors of lattice 1, and in Frank's approximation also of lattice 2, are the Burgers vectors, and the $\boldsymbol{x}^{(0)}$-points become the MS-points. Therefore, the theory of subgrain boundaries is a special case in the general theory of crystalline interfaces.

12.5. The $\boldsymbol{b}$-Lattice

The basic Eq. (12.3-6)

$$(\boldsymbol{I}-\boldsymbol{A}^{-1})\,\boldsymbol{x}^{(0)} = \boldsymbol{b}^{(L)}$$

can be interpreting as being an imaging relation of a b-lattice onto the two interpenetrating lattices 1 and 2, as was done in Section 11.2 in the representation of Frank's formula. The $\boldsymbol{b}^{(L)}$-vectors are defined as the translation vectors of lattice 1. *We now define the b-lattice as the lattice formed by all the possible translation vectors of lattice 1* when these are rearranged such that they have a common origin (Fig. 12.5/1). Thus, $\boldsymbol{b}^{(L)}$ stands for the lattice vector of the b-lattice. The b-lattice, from the point of view of geometry and orientation, is identical with the trans-

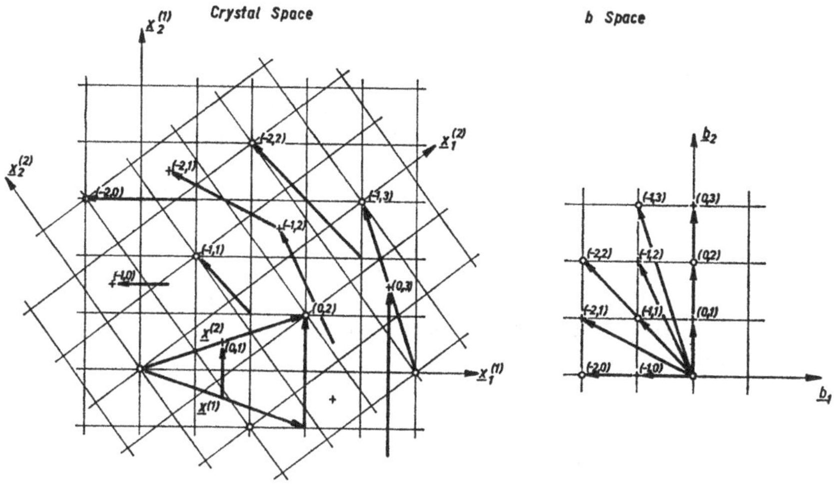

Fig. 12.5/1. The b-lattice pictured as the lattice of all the translation vectors of lattice 1

lation lattice 1. Therefore, the coordinate system of lattice 1 also fixes that of the b-lattice, and this, in turn, determines that of the 0-lattice.

We shall see later that this dualistic representation will be essential for the development of the theory.

12.6. Various Aspects of the 0-Points

According to the basic definition, the 0-points are positions where the two interpenetrating lattices show "best fit" (Section 12.1). This concept was specified in mathematical terms as the coincidences of elements of related equivalence classes (Section 12.3); in this form it is a generalization of the notion of lattice coincidence sites.

Furthermore, the 0-points proved to be generalized MS-points (solution of the quantized Frank's formula, Section 11.2). If we try to generalize Eq. (11.2-5) in order to obtain the basic Eq. (12.3-6), two problems arise:

1. In Frank's formula, $\boldsymbol{x}^{(1)} \approx \boldsymbol{x}^{(2)} \approx \boldsymbol{x}$ was not exactly fixed. The question now arises concerning the exact position of $\boldsymbol{x}$ for larger angles θ. ($\boldsymbol{x} = \boldsymbol{x}^{(1)}$, or $\boldsymbol{x}^{(2)}$, or some point in between?)

2. According to Frank's formula, $\boldsymbol{b}$ remains perpendicular to $\boldsymbol{x}$, and hence the b-coordinate system is perpendicular to that of the $\boldsymbol{x}$-vectors, where $\boldsymbol{b} = \boldsymbol{x}^{(2)} - \boldsymbol{x}^{(1)}$. This condition can only be maintained if $\boldsymbol{x}$ is an average value of $\boldsymbol{x}^{(1)}$ and $\boldsymbol{x}^{(2)}$. Otherwise the b- and the x-coordinate

systems are no longer perpendicular. Here the question arises concerning the meaning and orientation of the b-net.

Both questions can be answered if we assume that every lattice coincidence site is an MS-point. (The converse does not have to be true.) Comparing the dualism between the b- and the L-net given in Fig. 11.4/1 and the construction of the coincidence site lattice in Fig. 12.5/1, we see that $\boldsymbol{x} = \boldsymbol{x}^{(0)}$ has to be identified with $\boldsymbol{x}^{(2)}$ under the condition that

$$\boldsymbol{b} = \boldsymbol{x}^{(2)} - \boldsymbol{x}^{(1)} = \boldsymbol{b}^{(L)} \tag{12.6-1}$$

is a translation vector of lattice 1. The b-lattice is thus defined as the lattice of all the translation vectors of lattice 1 and the 0-lattice as the lattice of those points, $\boldsymbol{x}^{(2)}$ for which:

$$\boldsymbol{x}^{(2)} - \boldsymbol{x}^{(1)} = \boldsymbol{b}^{(L)}. \tag{12.6-2}$$

Together with Eq. (12.3-1) we obtain:

$$\boldsymbol{x}^{(0)} - \boldsymbol{A}^{-1}\,\boldsymbol{x}^{(0)} = \boldsymbol{b}^{(L)}$$

or

$$(\boldsymbol{I} - \boldsymbol{A}^{-1})\,\boldsymbol{x}^{(0)} = \boldsymbol{b}^{(L)} \qquad (12.6\text{-}3$$

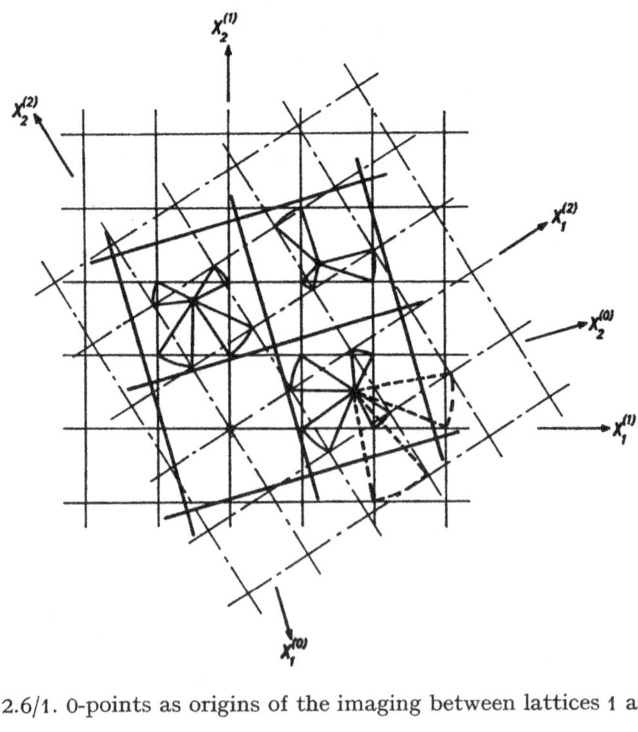

Fig. 12.6/1. 0-points as origins of the imaging between lattices 1 and 2

which is the basic Eq. (12.3-6). This is actually the way that the 0-lattice was found.

At this stage the 0-points, the generalized MS-points, were localized but their meaning was not immediately apparent; the meaning, however, is clear from Fig. 12.6/1 (Fig. 12.6/2 shows the complete pattern).

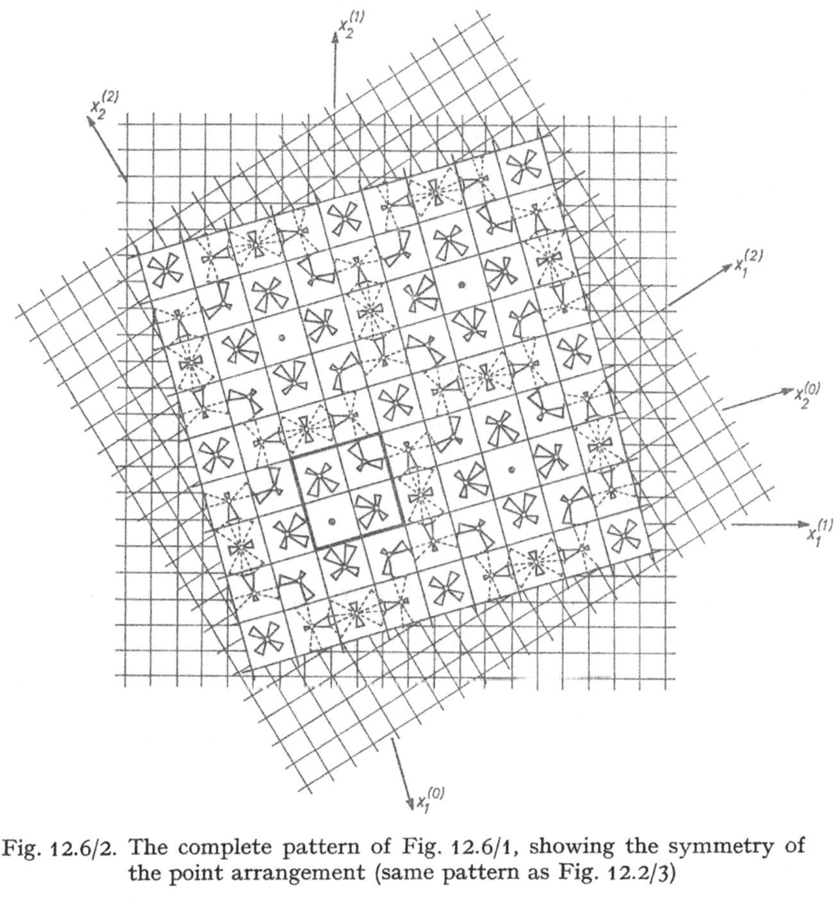

Fig. 12.6/2. The complete pattern of Fig. 12.6/1, showing the symmetry of the point arrangement (same pattern as Fig. 12.2/3)

By means of Eq. (12.1-4), lattice 2 was "created" out of lattice 1 by the transformation $\boldsymbol{A}$ where, for convenience, a lattice point was chosen as the origin. It can be seen that *the same lattice 2 in the same position can be created out of the same lattice 1 by the same transformation $\boldsymbol{A}$ but starting from other origins*. Naturally, on starting from a new origin, the pairing (the correlation) of the lattice points (of lattices 1 and 2) is changed. Formulated in an equation, this signifies a coordinate transformation by translation. Lattice 2 was defined by Eq. (12.2-1) ($\boldsymbol{x}^{(2L)} = \boldsymbol{A}\boldsymbol{x}^{(1L)}$). The new origin shall be $\boldsymbol{x}^{(0)}$, i.e.

$$\boldsymbol{x}^{(2L)} - \boldsymbol{x}^{(0)} = \boldsymbol{A}\,(\boldsymbol{x}^{(1L)} + \boldsymbol{b}^{(L)} - \boldsymbol{x}^{(0)})\,. \tag{12.6-4}$$

The point $\boldsymbol{x}^{(2L)}$ is here no longer related to $\boldsymbol{x}^{(1L)}$ but to another point of lattice 1 i.e. $\boldsymbol{x}^{(1L)} + \boldsymbol{b}^{(L)}$ (i.e. $\boldsymbol{x}^{(1L)} + \boldsymbol{t}^{(1)}$). From Eq. (12.6-4) it follows that

$$\boldsymbol{x}^{(2L)} - \boldsymbol{x}^{(0)} = \boldsymbol{A}\boldsymbol{x}^{(1L)} + \boldsymbol{A}\boldsymbol{b}^{(L)} - \boldsymbol{A}\boldsymbol{x}^{(0)} \tag{12.6-5}$$

and, applying Eq. (12.1-4)

$$-\boldsymbol{x}^{(0)} = \boldsymbol{A}\boldsymbol{b}^{(L)} - \boldsymbol{A}\boldsymbol{x}^{(0)}\,. \tag{12.6-6}$$

Upon multiplication of both sides by $\boldsymbol{A}^{-1}$, the basic Eq. (12.3-6) is produced. Therefore, all the 0-points are origins of the type described above.

Naturally, the various meanings of the 0-points indicated in this section, such as:

Coincidences of equivalent points,

Generalization of MS-points, and

Origins for the relation between lattice 1 and lattice 2

are different manifestations of the same situation. However, which of the various aspects has to be emphasized will depend on the problem being treated.

12.7. The Actual Formulation of the Transformation *A*

We start with the general case where two crystals of different structures in a given relative orientation are joined. We assume that the lattice constants a, b, c, α, β, γ (Section 2.2) of the two lattices are known. Starting from a cubic lattice of unit period 1 Å we "produce" lattice 1 by the transformation $\boldsymbol{S}^{(1)}$

$$\boldsymbol{x}^{(1)} = \boldsymbol{S}^{(1)}\,\boldsymbol{x}^{(\text{orth})}\,. \tag{12.7-1}$$

$\boldsymbol{S}^{(1)}$ is formulated according to the general method of Section 2.2. Of the nine coefficients, three can be chosen by fixing the orientation of the unit cell (e.g. $\boldsymbol{c}$ in z-axis and $\boldsymbol{a}$ in (xz)-plane]. The detailed procedure is given in (Eq. A1-26). Lattice 2 is produced in the same manner, but in order to be free to choose its orientation we rotate it by $\boldsymbol{R}$. Hence:

$$\boldsymbol{x}^{(2)} = \boldsymbol{R}\boldsymbol{S}^{(2)}\,\boldsymbol{x}^{(\text{orth})}\,. \tag{12.7-2}$$

By eliminating $\boldsymbol{x}^{(\text{orth})}$, we obtain

$$\boldsymbol{x}^{(2)} = \{\boldsymbol{R}\boldsymbol{S}^{(2)}\boldsymbol{S}^{(1)-1}\}\,\boldsymbol{x}^{(1)} := \boldsymbol{A}\boldsymbol{x}^{(1)}\,. \tag{12.7-3}$$

$\boldsymbol{A}$ is composed of three different transformations. Eq. (12.7-3) is valid in the orthogonal coordinate system. From this follows:

$$(\boldsymbol{I}-\boldsymbol{A}^{-1})=(\boldsymbol{I}-\boldsymbol{S}^{(1)}(\boldsymbol{S}^{(2)})^{-1}\boldsymbol{R}^{-1}) \tag{12.7-4}$$

and, if $|\boldsymbol{I}-\boldsymbol{A}^{-1}|\neq 0$, according to Eq. (12.3-7):

$$\boldsymbol{x}^{(0)}=(\boldsymbol{I}-\boldsymbol{S}^{(1)}(\boldsymbol{S}^{(2)})^{-1}\boldsymbol{R}^{-1})^{-1}\,\boldsymbol{b}^{(L)}. \tag{12.7-5}$$

As this equation is formulated in the orthogonal coordinate system, the unit vectors of the b-lattice are the column vectors of the matrix $\boldsymbol{S}^{(1)}$. In this manner, the unit vectors of the 0-lattice are determined as images of the b-unit vectors.

If both crystals are of the same kind, i.e. if we deal with grain boundaries, $\boldsymbol{S}^{(2)}=\boldsymbol{S}^{(1)}:=\boldsymbol{S}$, and hence $\boldsymbol{A}=\boldsymbol{R}$. The $\boldsymbol{S}$-matrix then only plays a role in the determination of the $\boldsymbol{b}^{(L)}$-vectors.

It will become useful to express all the vectors in the crystal coordinate system (of lattice 1). Then, every matrix $\boldsymbol{M}$ is transformed from the orthogonal- to the crystal coordinates (Index C) by:

$$\boldsymbol{M}^{(C)}=(\boldsymbol{S}^{(1)})^{-1}\boldsymbol{M}\boldsymbol{S}^{(1)} \tag{12.7-6}$$

and all the vectors $\boldsymbol{x}$ by:

$$\boldsymbol{x}^{(C)}=(\boldsymbol{S}^{(1)})^{-1}\,\boldsymbol{x}. \tag{12.7-7}$$

As we deal here with a coordinate transformation and not with a vector transformation, the $\boldsymbol{x}$ in Eq. (12.7-7) has the same significance as $\boldsymbol{\xi}$ in Eq. (2.2-5).

In the crystal coordinate system $\boldsymbol{A}$ becomes:

$$\boldsymbol{A}^{(C)}=(\boldsymbol{S}^{(1)})^{-1}\{\boldsymbol{R}\boldsymbol{S}^{(2)}(\boldsymbol{S}^{(1)})^{-1}\}\boldsymbol{S}^{(1)}=(\boldsymbol{S}^{(1)})^{-1}\boldsymbol{R}\boldsymbol{S}^{(2)} \tag{12.7-8}$$

and the basic equation:

$$\{(\boldsymbol{S}^{(1)})^{-1}[\boldsymbol{I}-\boldsymbol{S}^{(1)}(\boldsymbol{S}^{(2)})^{-1}\boldsymbol{R}^{-1}]\boldsymbol{S}^{(1)}\}\{(\boldsymbol{S}^{(1)})^{-1}\boldsymbol{x}^{(0)}\}=\{(\boldsymbol{S}^{(1)})^{-1}\boldsymbol{b}^{(L)}\}$$

i.e.

$$(\boldsymbol{I}-(\boldsymbol{S}^{(2)})^{-1}\boldsymbol{R}^{-1}\boldsymbol{S}^{(1)})\,\boldsymbol{x}^{(0)(C)}=\boldsymbol{b}^{(L)(C)}. \tag{12.7-9}$$

As an example, we investigate the rotation on the $(1\bar{1}0)$-plane in the bcc structure, which is illustrated by moiré model No. 1. The unit vectors of the crystal coordinate system are given by the vectors [1/2, 1/2, 1/2] and [001] in the cubic system (Fig. 12.7/1), hence

$$\boldsymbol{S}=\begin{pmatrix}\frac{\sqrt{2}}{2} & 0\\ \frac{1}{2} & 1\end{pmatrix}. \tag{12.7-10}$$

The unit vectors of the 0-lattice in orthogonal coordinates follow from the matrix (12.3-12) multiplied by the column vectors of $\boldsymbol{S}$ [Eq. (12.7-10)].

$$\begin{aligned} \boldsymbol{u}_1^{(0)} &= [(1/4)(\sqrt{2}+\cot(\theta/2)),\ (1/4)(1-\sqrt{2}\cot(\theta/2))] \\ \boldsymbol{u}_2^{(0)} &= [(1/2)\cot(\theta/2),\ 1/2]. \end{aligned} \tag{12.7-11}$$

In crystal coordinates, these vectors become:

$$\begin{aligned} \boldsymbol{u}_1'^{(0)(C)} &= [(1/2)(1+(\sqrt{2}/2)\cot(\theta/2)),\ -(3\sqrt{2}/8)\cot(\theta/2)] \\ \boldsymbol{u}_2'^{(0)(C)} &= [(\sqrt{2}/2)\cot(\theta/2),\ (1/2)(1-(\sqrt{2}/2)\cot(\theta/2)]. \end{aligned} \tag{12.7-12}$$

This result is obtained either by formulating the transformation $\boldsymbol{A}$ in crystal coordinates

$$\boldsymbol{A}^{(C)} = \boldsymbol{S}^{-1}\boldsymbol{R}\boldsymbol{S} \tag{12.7-13}$$

or by transforming the matrix $(\boldsymbol{I}-\boldsymbol{A}^{-1})^{-1}$. The column vectors of the transformed matrix are now the unit vectors of the 0-lattice because, in the crystal coordinate system, the unit-$\boldsymbol{b}^{(L)}$-vectors are of the form $\langle 1, 0\rangle$. As a third possibility, the orthogonal $\boldsymbol{u}_i^{(0)}$ can be transformed by Eq. (12.7-7).

The procedure which was given in this section is useful for relatively small rotations. For large rotations an additional point of view appears which will be dealt with in the next section.

12.8. Choice between Various Possible Transformations $\boldsymbol{A}$

In Section 12.7 the example of rotation on the $\{110\}$-plane in the bcc structure was calculated. Fig. 12.8/1 shows a comparison between the calculated 0-lattice and the corresponding moiré model (model No. 1). For a small rotation the agreement is good; for large rotations, however, the moiré model shows a configuration which is completely different from the pattern calculated on the basis of the transformation $\boldsymbol{A}=\boldsymbol{R}$ [Eq. (12.3-8)]. The reason for this discrepancy lies in the fact that, on increasing the angle of rotation above a certain value, $\boldsymbol{R}$ no longer describes the relation between the nearest neighbours in the two lattices. Visually, however, the nearest neighbours are automatically related in the moiré pattern. Also, from the physical standpoint the relation between nearest neighbours is of primary importance.

In order to correlate the nearest neighbours for larger rotations the following vector relations (in coordinates of their respective crystals) must exist:

$$\boldsymbol{x}^{(1)}[1, 0] \quad \text{to} \quad \boldsymbol{x}^{(2)}[1, -1]$$

and

$$\boldsymbol{x}^{(1)}[0, 1] \quad \text{to} \quad \boldsymbol{x}^{(2)}[1, 0].$$

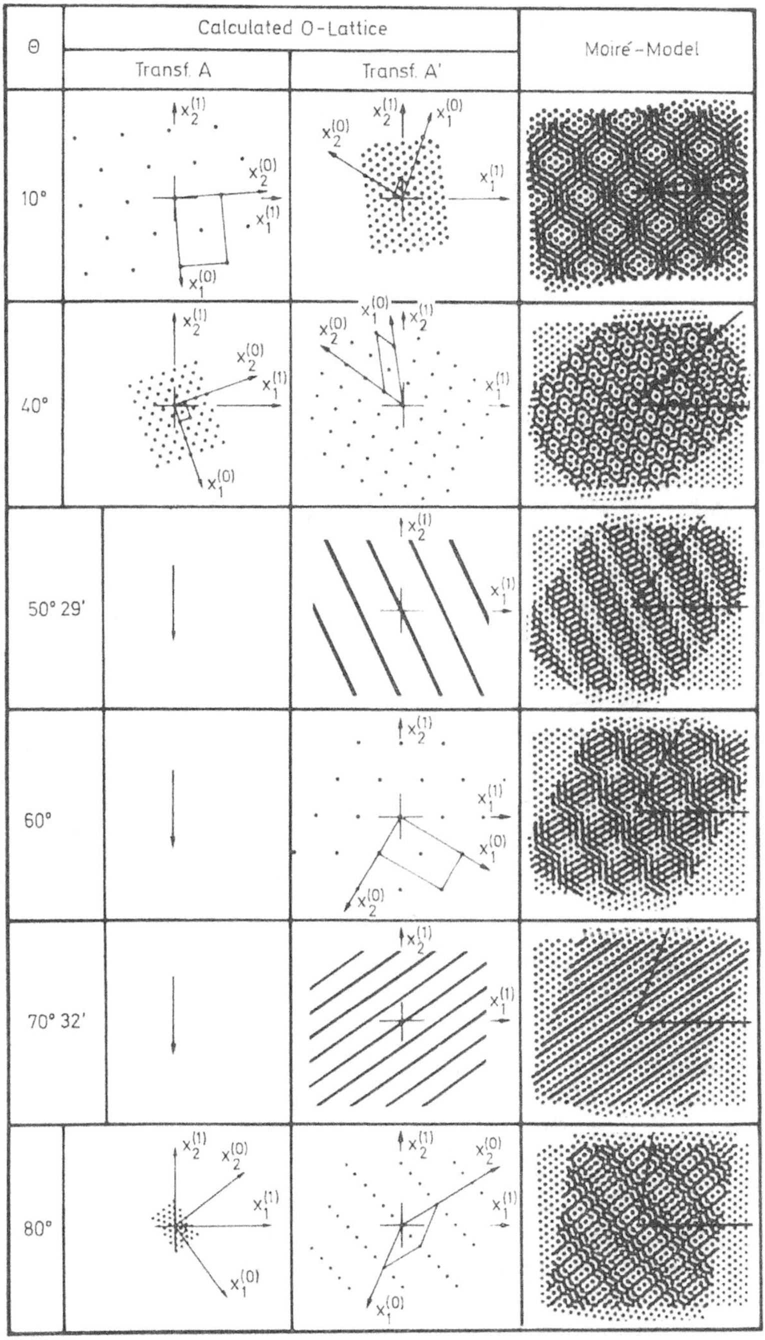

Fig. 12.8/1. Comparison between calculated 0-lattices and the moiré models No. 1. [$\boldsymbol{A}$: Eq. (12.7-3), $\boldsymbol{A}'$: Eq. (12.8-9)]

First $\boldsymbol{x}^{(1)}$ is transformed into $\boldsymbol{x}^{(1)'}$ in lattice 1 by the transformation $\boldsymbol{U}$ and from there into $\boldsymbol{x}^{(2)}$ by the rotation $\boldsymbol{R}$ (Fig. 12.8/2).

$$\boldsymbol{x}^{(1)'} = \boldsymbol{U}\boldsymbol{x}^{(1)} \tag{12.8-1}$$

$$\boldsymbol{U} = \begin{pmatrix} 1 & 1 \\ -1 & 0 \end{pmatrix}. \tag{12.8-2}$$

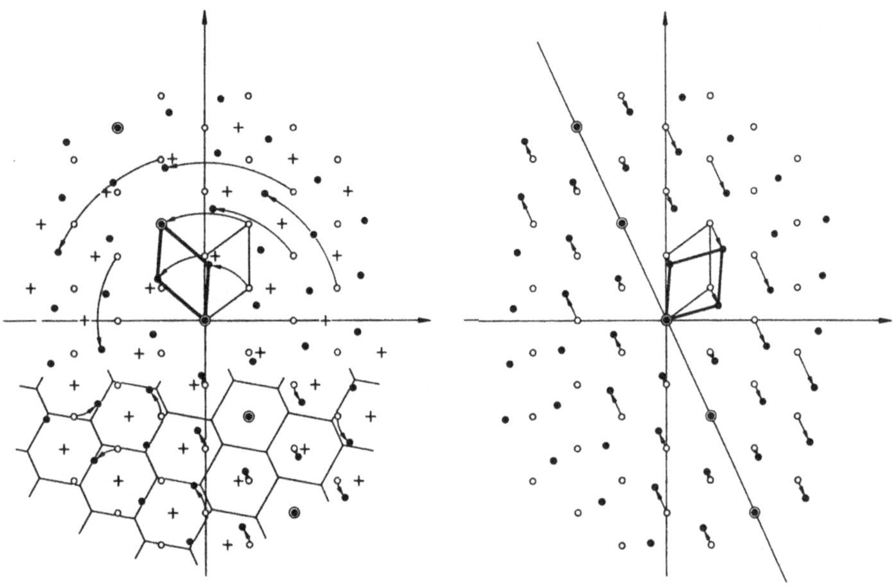

a) b)

Fig. 12.8/2a and b. Point pattern for $\theta = 50°\,29'$ 0-lattices for: a) $\boldsymbol{A}$ = rotation, Eq. (12.3-8), b) $\boldsymbol{A}'$ = pure shear, Eq. (12.8-9)

This means that we choose new unit vectors in crystal 1, and starting from these, we apply the rotation $\boldsymbol{S}^{-1}\boldsymbol{R}\boldsymbol{S}$ (in crystal coordinates) in order to arrive at $\boldsymbol{x}^{(2)}$.

$$\boldsymbol{x}^{(2)} = (\boldsymbol{S}^{-1}\boldsymbol{R}\boldsymbol{S}\boldsymbol{U})\,\boldsymbol{x}^{(1)} := \boldsymbol{A}'^{(C)}\,\boldsymbol{x}^{(1)}. \tag{12.8-3}$$

$\boldsymbol{U}$ is a "unimodular" transformation, i.e. a transformation which leaves the volume of the unit cell unchanged ($\det(\boldsymbol{U}) = 1$) although it may, however, change its shape. The orthogonal transformations, and especially the symmetry operations, which leave the volume as well as the shape unchanged, form a subgroup of the unimodular transformations and these, in turn, form a subgroup of the general linear transformations.

Referred to the orthogonal coordinate system (Fig. 12.8/3):

$$\boldsymbol{A}' = \boldsymbol{S}\boldsymbol{A}^{(C)}\boldsymbol{S}^{-1} = \boldsymbol{S}(\boldsymbol{S}^{-1}\boldsymbol{R}\boldsymbol{S}\boldsymbol{U})\boldsymbol{S}^{-1} = \boldsymbol{R}\boldsymbol{S}\boldsymbol{U}\boldsymbol{S}^{-1}. \tag{12.8-4}$$

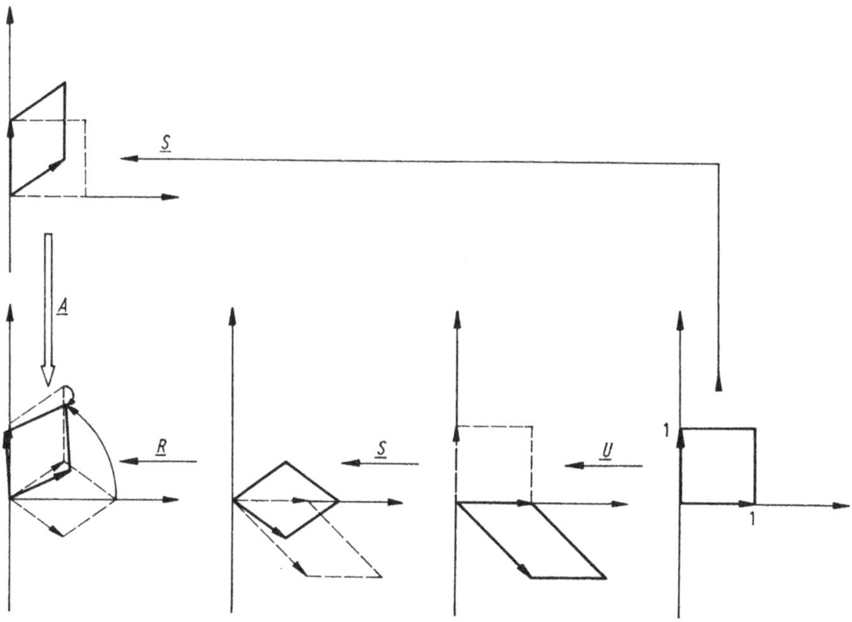

Fig. 12.8/3. Construction of the transformation $\boldsymbol{A}'$ [Eq. (12.8-4)]

$\boldsymbol{A}'\boldsymbol{S} = \boldsymbol{R}\boldsymbol{S}\boldsymbol{U} \rightarrow \boldsymbol{A}' = \boldsymbol{R}\boldsymbol{S}\boldsymbol{U}\boldsymbol{S}^{-1}$

The case of the small rotation is described by $\boldsymbol{U} = \boldsymbol{I}$. Then, in crystal coordinates, Eq. (12.8-3) corresponds to (12.7-8) (with $\boldsymbol{S}^{(2)} = \boldsymbol{S}^{(1)} = \boldsymbol{S}$) and in orthogonal coordinates (12.8-4) to (12.3-8).

In crystal coordinates

$$(\boldsymbol{I} - \boldsymbol{A}'^{(C)-1}) = (\boldsymbol{I} - \boldsymbol{U}^{-1}\boldsymbol{S}^{-1}\boldsymbol{R}^{-1}\boldsymbol{S}) \tag{12.8-5}$$

and in orthogonal coordinates

$$(\boldsymbol{I} - \boldsymbol{A}'^{-1}) = (\boldsymbol{I} - \boldsymbol{S}\boldsymbol{U}^{-1}\boldsymbol{S}^{-1}\boldsymbol{R}^{-1}). \tag{12.8-6}$$

Explicitly the matrices are:

$$A'^{(C)} = \begin{pmatrix} \cos\theta + (\sqrt{2}/2)\sin\theta & \cos\theta - (\sqrt{2}/2)\sin\theta \\ -\cos\theta + (\sqrt{2}/4)\sin\theta & (3\sqrt{2}/4)\sin\theta \end{pmatrix} \tag{12.8-7}$$

$$(\boldsymbol{I} - (\boldsymbol{A}'^{(C)})^{-1}) = \begin{pmatrix} 1 - (3\sqrt{2}/4)\sin\theta & \cos\theta - (\sqrt{2}/2)\sin\theta \\ -\cos\theta + (\sqrt{2}/4)\sin\theta & 1 - \cos\theta - (\sqrt{2}/2)\sin\theta \end{pmatrix} \tag{12.8-8}$$

and in orthogonal coordinates

$$A' = \begin{pmatrix} (1/2)\cos\theta + (3\sqrt{2}/4)\sin\theta & (\sqrt{2}/2)\cos\theta - (1/2)\sin\theta \\ -(3\sqrt{2}/4)\cos\theta + (1/2)\sin\theta & (1/2)\cos\theta + (\sqrt{2}/2)\sin\theta \end{pmatrix} \quad (12.8\text{-}9)$$

and

$$(I - A'^{-1}) =$$

$$\begin{pmatrix} 1-(1/2)\cos\theta - (\sqrt{2}/2)\sin\theta & (\sqrt{2}/2)\cos\theta - (1/2)\sin\theta \\ -(3\sqrt{2}/4)\cos\theta + (1/2)\sin\theta & 1-(1/2)\cos\theta - (3\sqrt{2}/4)\sin\theta \end{pmatrix}. \quad (12.8\text{-}10)$$

The determinants, as volume ratios of related unit cells, are independent of the coordinate system of the transformation. Since A' is a rotation multiplied by a unimodular transformation, it follows that

$$\det(A') = 1. \quad (12.8\text{-}11)$$

However, the determinant

$$|I - A'^{-1}| = 2 - \cos\theta - (5\sqrt{2}/4)\sin\theta \quad (12.8\text{-}12)$$

while the simple rotation:

$$|I - A^{-1}| = 2(1 - \cos\theta). \quad (12.8\text{-}13)$$

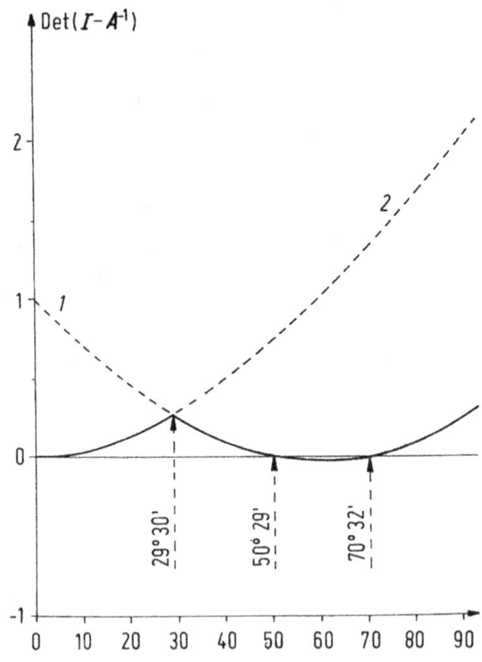

Fig. 12.8/4. Determinant $|I - A^{-1}|$. *1.* Eq. (12.8-12), *2.* Eq. (12.8-13)

Both determinants are indicated in Fig. 12.8/4. The determinant $|\boldsymbol{I}-\boldsymbol{A}^{-1}|$ is the ratio of the volume of the b-unit cell (in this case the surface) to the 0-lattice unit cell. The smaller this determinant, the larger becomes the unit cell of the 0-lattice and the closer are the correlated points in lattices 1 and 2.

The nearest neighbours in both lattices are related by that transformation $\boldsymbol{A}$ *selected from various possibilities for which the determinant* $|\boldsymbol{I}-\boldsymbol{A}^{-1}|$ *acquires the smallest absolute value.*

From Fig. 12.8/4 we see that for θ less than 29° 30′, the transformation $\boldsymbol{A}$ relates the nearest neighbours, and above this value the relation is given by the transformation $\boldsymbol{A}'$, which contains $\boldsymbol{U}$. Fig. 12.8/2 shows that in the two crystals which are of identical structure, *different unit cells* have to be chosen in order to relate the nearest neighbours.

Fig. 12.8/5 gives the determinant $|\boldsymbol{I}-\boldsymbol{A}'^{-1}|$ for a rotation on the {1 1 2}-plane. In this case four different transformations are required for a rotation of more than 90°. This example may be studied on the moiré model No. 2 (see also the Table 13.3/T 1).

In the general case where two different crystals have to be fitted together the procedure corresponds to that given in Section 12.7. Starting

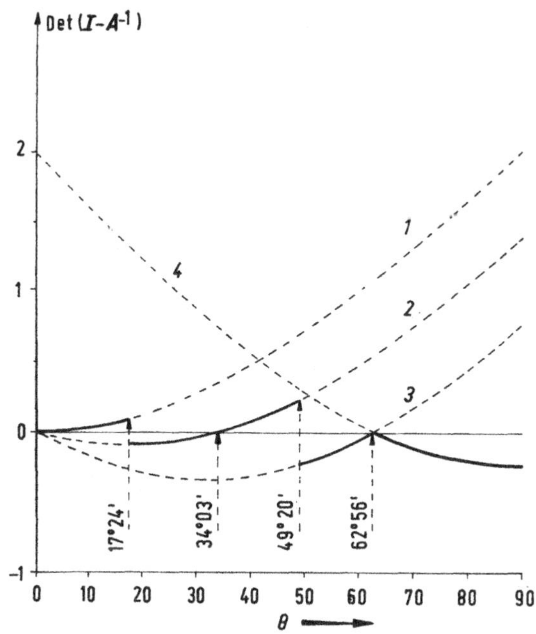

Fig. 12.8/5. Determinant $|\boldsymbol{I}-\boldsymbol{A}^{-1}|$ for rotation on the {1 1 2}-plane

from an orthogonal unit cell

$$\boldsymbol{x}^{(1)} = \boldsymbol{S}^{(1)} \boldsymbol{x}^{(\text{orth})}. \tag{12.8-14}$$

In order to determine $\boldsymbol{x}^{(2)}$ it is, in general, more convenient to first redefine the coordinate system of the orthogonal unit cell by $\boldsymbol{U}$ e.g. Eq. (12.8-2) before transforming that into the unit cell of lattice 2.

$$\boldsymbol{x}^{(2)} = \boldsymbol{R}\boldsymbol{S}^{(2)} \boldsymbol{U}\boldsymbol{x}^{(\text{orth})}. \tag{12.8-15}$$

Hence, in orthogonal coordinates

$$\boldsymbol{A}' = \boldsymbol{R}\boldsymbol{S}^{(2)} \boldsymbol{U}\boldsymbol{S}^{(1)-1}. \tag{12.8-16}$$

For crystals of the same type $\boldsymbol{S}^{(2)} = \boldsymbol{S}^{(1)} = \boldsymbol{S}$ and $\boldsymbol{A}'$ becomes Eq. (12.8-4).

$\boldsymbol{x}^{(2)}$ could also be represented by

$$\boldsymbol{x}^{(2)} = \boldsymbol{R}\boldsymbol{U}'\boldsymbol{S}^{(2)} \boldsymbol{x}^{(\text{orth})} \tag{12.8-17}$$

and

$$\boldsymbol{A}' = \boldsymbol{R}\boldsymbol{U}'\boldsymbol{S}^{(2)} \boldsymbol{S}^{(1)-1} \tag{12.8-18}$$

but here $\boldsymbol{U}'$ would have the less convenient form

$$\boldsymbol{U}' = \begin{pmatrix} 1/2 & \sqrt{2}/2 \\ -3\sqrt{2}/4 & 1/2 \end{pmatrix} \tag{12.8-19}$$

instead of Eq. (12.8-2).

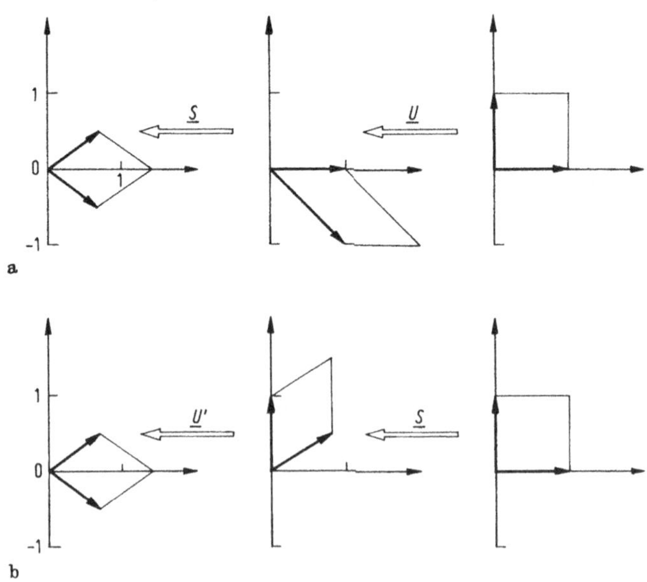

Fig. 12.8/6. Difference in the meaning of $\boldsymbol{U}$, Eq. (12.8-2), and $\boldsymbol{U}'$, Eq. (12.8-19)

The difference in the operation of $\boldsymbol{U}$ and $\boldsymbol{U}'$ is shown in Fig. 12.8/6.

Summarizing, the transformation $\boldsymbol{A}$ can be expressed in its general form either by Eq. (12.8-16) or by (12.8-18).

It is often the case that relations between the two lattices may be known which do not relate the nearest neighbours, e.g. twin laws are described as a rotation by 180° around a twin axis or as mirror images across certain planes. We now present a method which allows to arrive at the transformation $\boldsymbol{A}$, indicating the nearest neighbour relations, while starting from an arbitrary point to point relation $\bar{\boldsymbol{A}}$ between the two lattices. The calculation is carried out in the crystal coordinate system.

We first transform a unit vector of lattice 1, e.g. $\boldsymbol{u}_1^{(1)}$ and arrive at a lattice vector of lattice 2, i.e. $\boldsymbol{x}^{(2L)}$

$$\boldsymbol{x}^{(2L)} = \bar{\boldsymbol{A}}\boldsymbol{u}_i^{(1)}. \tag{12.8-20}$$

Then, by another transformation $\boldsymbol{U}$ we transform $\boldsymbol{u}_i^{(1)}$ into that translation vector $\boldsymbol{t}_i^{(1)}$ (of lattice 1) which lies closest to $\boldsymbol{x}^{(2L)}$:

$$\boldsymbol{t}_i^{(1)} = \boldsymbol{U}\boldsymbol{u}_i^{(1)}. \tag{12.8-21}$$

The transformation which relates $\boldsymbol{t}_i^{(1)}$ to $\boldsymbol{x}^{(2L)}$ is the transformation $\boldsymbol{A}$ relating nearest neighbours.

$$\boldsymbol{x}^{(2L)} = (\bar{\boldsymbol{A}}\boldsymbol{U}^{-1})\,\boldsymbol{t}_i^{(1)} := \boldsymbol{A}\boldsymbol{x}^{(1L)} \tag{12.8-22}$$

and

$$(\boldsymbol{I} - \boldsymbol{A}^{-1}) = (\boldsymbol{I} - \boldsymbol{U}\bar{\boldsymbol{A}}^{-1}). \tag{12.8-23}$$

The problem is now to find the transformation $\boldsymbol{U}$. For this purpose we determine that unit cell of lattice 1 which contains $\boldsymbol{x}^{(2L)}$, i.e. we separate the external- and internal coordinates of $\boldsymbol{x}^{(2)}$ (Section 12.3). We decompose $\boldsymbol{x}^{(2L)}$ into $\boldsymbol{t}_0$ and $\boldsymbol{x}$.

$$\boldsymbol{x}^{(2L)} = \boldsymbol{t}_0 + \boldsymbol{x}. \tag{12.8-24}$$

(This is the reason for working in the crystal coordinate system.)

Now we have to find that corner of this unit cell which is nearest to $\boldsymbol{x}^{(2L)}$. We subdivide the unit cell into Wigner-Seitz cells (W-S-cells) (Section 2.2) around every corner and look for that W-S-cell which contains $\boldsymbol{x}^{(2L)}$.

In order to solve this problem we first have to find the equation of the W-S-cell wall between two arbitrary points $\boldsymbol{t}'$ and $\boldsymbol{t}$ (Fig. 12.8/7). Initially, we formulate the problem in orthogonal coordinates. The projection of the vector $\boldsymbol{x}$ onto a vector through the origin parallel to $\boldsymbol{t}' - \boldsymbol{t}$ must be equal to the projection of the vector $\boldsymbol{t} + (1/2)(\boldsymbol{t}' - \boldsymbol{t})$, i.e.

$$\left(\frac{\boldsymbol{t}' - \boldsymbol{t}}{|\boldsymbol{t}' - \boldsymbol{t}|} \cdot \boldsymbol{x}\right) = \left[\left(\boldsymbol{t} + \frac{\boldsymbol{t}' - \boldsymbol{t}}{2}\right) \cdot \frac{\boldsymbol{t}' - \boldsymbol{t}}{|\boldsymbol{t}' - \boldsymbol{t}|}\right] \tag{12.8-25}$$

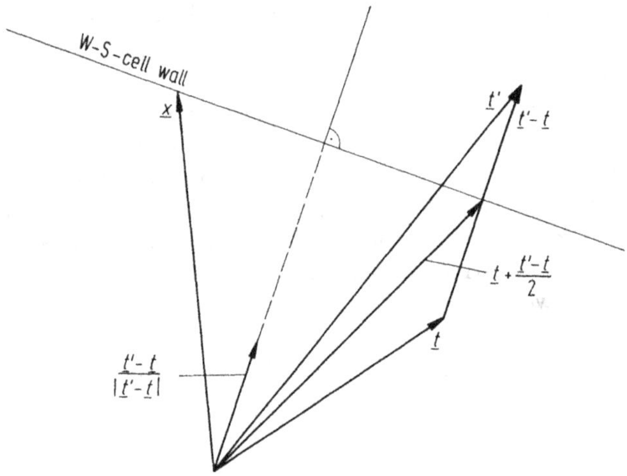

Fig. 12.8/7. Elements for the calculation of the Wigner-Seitz cell walls

then

$$(\boldsymbol{t}'-\boldsymbol{t})\cdot\boldsymbol{x}-(1/2)(\boldsymbol{t}'+\boldsymbol{t})\cdot(\boldsymbol{t}'-\boldsymbol{t})=0 \qquad (12.8\text{-}26)$$

and

$$[(\boldsymbol{t}'\cdot\boldsymbol{x})-(1/2)(\boldsymbol{t}'\cdot\boldsymbol{t}')]-[(\boldsymbol{t}\cdot\boldsymbol{x})-(1/2)(\boldsymbol{t}\cdot\boldsymbol{t})]=0. \qquad (12.8\text{-}27)$$

This is the equation of the W-S-wall. We abbreviate it to

$$f(\boldsymbol{x},\boldsymbol{t})'-f(\boldsymbol{x},\boldsymbol{t})=0 \qquad (12.8\text{-}28)$$

where

$$f(\boldsymbol{x},\boldsymbol{t})=(\boldsymbol{t}\cdot\boldsymbol{x})-(1/2)(\boldsymbol{t}\cdot\boldsymbol{t}). \qquad (12.8\text{-}29)$$

In general coordinates $f(\boldsymbol{x},\boldsymbol{t})$ becomes:

$$f(\boldsymbol{x},\boldsymbol{t})=(\boldsymbol{t}^T\boldsymbol{G}\boldsymbol{x})-(1/2)(\boldsymbol{t}^T\boldsymbol{G}\boldsymbol{t}). \qquad (12.8\text{-}30)$$

Hence, we can separate the equation of the W-S-wall (placed between two points $\boldsymbol{t}'$ and $\boldsymbol{t}$) into two similar expressions, which individually depend on only one of the points.

If we choose a point $\boldsymbol{x}$ *not* on the Wigner-Seitz wall but on that side of the wall on which $\boldsymbol{t}'$ lies, then the expression (12.8-28) becomes positive, i.e. $f(\boldsymbol{x},\boldsymbol{t}')>f(\boldsymbol{x},\boldsymbol{t})$, and vice versa if $\boldsymbol{x}$ lies nearer to $\boldsymbol{t}$.

This leads to the procedure for determining the corner of the unit cell which is nearest to $\boldsymbol{x}^{(2L)}$. We have to determine $f(\boldsymbol{x},\boldsymbol{t})$ for all the following $\boldsymbol{t}$-vectors: [100], [010], [110], [001], [101], [011] and [111], while for $\boldsymbol{x}$ the internal coordinates of $\boldsymbol{x}^{(2L)}$ are chosen. That $\boldsymbol{t}$ for which $f(\boldsymbol{x},\boldsymbol{t})$ has the largest *positive* value is the one which is nearest to $\boldsymbol{x}$. The sum of $\boldsymbol{t}^{(1)}$ is given by the $\boldsymbol{t}$ chosen in the above manner together with the external coordinates of $\boldsymbol{x}^{(2L)}$.

$$\boldsymbol{t}^{(1)} = \boldsymbol{t}_0 + \boldsymbol{t}. \tag{12.8-31}$$

If $\boldsymbol{u}_i^{(1)}$ is $\boldsymbol{u}_1^{(1)}$, then $\boldsymbol{t}^{(1)}$ is the first column vector of $\boldsymbol{U}$ (because in the crystal coordinate system, $\boldsymbol{u}_1^{(1)} = [1\,0\,0]$). The second and the third column vectors of $\boldsymbol{U}$ are determined in the same way, with $\boldsymbol{u}_2^{(1)}$ and $\boldsymbol{u}_3^{(1)}$.

If two of the values $f(\boldsymbol{x}, \boldsymbol{t})$ are the same, then the point $\boldsymbol{x}^{(2L)}$ lies at the same distance from two different points. (This situation corresponds, for example, to the meeting point of the curves in Fig. 12.8/3). If all the values of $f(\boldsymbol{x}, \boldsymbol{t})$ are ≤ 0, then $\boldsymbol{t} = [0\,0\,0]$ and $\boldsymbol{t}^{(1)} = \boldsymbol{t}_0$. If several $\boldsymbol{t}$-vectors are closest to $\boldsymbol{x}$, $\boldsymbol{t}$ should be chosen such that $\boldsymbol{U}$ is unimodular ($\det(\boldsymbol{U}) = 1$), as only then the two lattices are in a 1 to 1 relation.

This procedure can be programmed on a computer.

An equivalent procedure is to find the smallest value of the square of the distance to the corner t:

$$(\boldsymbol{t} - \boldsymbol{x})^T \boldsymbol{G} (\boldsymbol{t} - \boldsymbol{x}) \tag{12.8-32}$$

where for $\boldsymbol{t}$ all the eight corners of the unit cell have to be checked.

12.9. Solutions of the Basic Equation

After having formulated the transformation $\boldsymbol{A}$ in Section 12.8 we now discuss the solutions of the basic equation.

$$(\boldsymbol{I} - \boldsymbol{A}^{-1})\, \boldsymbol{x}^{(0)} = \boldsymbol{b}^{(L)}. \tag{12.9-1}$$

If we set

$$(\boldsymbol{I} - \boldsymbol{A}^{-1}) := \boldsymbol{T} \tag{12.9-2}$$

Eq. (12.9-1) may then be written as

$$\boldsymbol{T} \boldsymbol{x}^{(0)} = \boldsymbol{b}^{(L)} \tag{12.9-3}$$

or, in index notation:

$$t_k^i\, x^{k\,(0)} = b^{i\,(L)} \tag{12.9-4}$$

and in the expanded form:

$$\begin{aligned} t_1^1\, x^{1\,(0)} + t_2^1\, x^{2\,(0)} + t_3^1\, x^{3\,(0)} &= b^{1\,(L)} \\ t_1^2\, x^{1\,(0)} + t_2^2\, x^{2\,(0)} + t_3^2\, x^{3\,(0)} &= b^{2\,(L)} \\ t_1^3\, x^{1\,(0)} + t_2^3\, x^{2\,(0)} + t_3^3\, x^{3\,(0)} &= b^{3\,(L)}. \end{aligned} \tag{12.9-5}$$

The rank of the matrix $\boldsymbol{T}$ may be 3, 2, 1 or 0 in the (trivial) limiting case.

In the case where the rank $(\boldsymbol{T}) = 3$ the solutions $\boldsymbol{x}^{(0)}$ are given by

$$\boldsymbol{x}^{(0)} = \boldsymbol{T}^{-1}\, \boldsymbol{b}^{(L)}. \tag{12.9-6}$$

In crystal coordinates the $\boldsymbol{b}^{(L)}$-vectors are of the form $\langle 1\,0\,0 \rangle$, and hence, the column vectors of the matrix $\boldsymbol{T}^{-1}$ are the unit vectors of the 0-lattice.

In the examples of rank$(\boldsymbol{T})<3$ the Eqs. (12.9-5) may be solved by standard linear algebraic methods. Here we present methods which are specially suited to the solution of our problems involving three, two or one dimension.

The Eq. (12.9-5) can for given $b^{i(L)}$ be understood as equations of three planes. In the case of rank$(\boldsymbol{T})=3$ the three normal vectors on these planes are linearly independent and the planes intersect in a point, which represents the solution. The normal vectors on the planes are the row vectors of the matrix $\boldsymbol{T}$, i.e. (t_1^i, t_2^i, t_3^i). If the equations are formulated in the crystal coordinate system, then the coordinates of these row vectors are referred to the reciprocal lattice related to the crystal lattice (since here the lower indices vary). (In the orthogonal system the crystal space and the reciprocal space are of the same structure, but of different dimensions $\text{Å} \leftrightarrow \text{Å}^{-1}$.)

In the case of rank$(\boldsymbol{T})=2$ the three planes are parallel to a straight line, and thus the three normal vectors lie in a plane. For arbitrary values of $b^{i(L)}$, the three planes would intersect in pairs in three parallel lines and the three equations would not have a common solution. A change of $b^{i(L)}$ means a translation of the corresponding plane. In order to be able to solve Eq. (12.9-5) the three planes must intersect in *one* line. In this manner, if two b-coordinates are given the third is also fixed. Assuming that b^1 and b^2 are given, b^3 has to be of the form

$$b^3=\lambda_1 b^1+\lambda_2 b^2 \tag{12.9-7}$$

or

$$c_1 b^1+c_2 b^2+c_3 b^3=0. \tag{12.9-8}$$

For the moment we consider all b^i as continuous variables. Eq. (12.9-8) represents a *plane in the b-space* passing through the origin, and we term this b-plane a two-dimensional *b-subspace*. This means that for rank$(\boldsymbol{T})=2$, solutions to Eq. (12.9-5) can only be found for b-lattice points which lie in that b-subspace.

The solutions, $\boldsymbol{x}^{(0)}$ of Eq. (12.9-5) are no longer single points, but are now all the points on the common line of intersection of the three planes. For every $\boldsymbol{b}^{(L)}$ in the two-dimensional b-subspace there exists a straight $x^{(0)}$-line (or shorter 0-*line*). We already came across such a situation in the interpretation of Frank's formula (Section 11.2), where the b-subspace was the plane $b^3=0$ and the 0-lines were straight lines parallel to the z-axis of rotation.

On solving Eq. (12.9-5) we not only have to determine the 0-lines but also the normal to the b-subspace c_i, [Eq. (12.9-8)]. First we look for a vector through the origin parallel to the 0-lines. Once this vector is known, the positions of the 0-lines are determined by the values of $\boldsymbol{b}^{(L)}$.

As was mentioned above, the 0-lines are the lines of intersection of three planes. The lines of intersection are perpendicular to the three normal vectors [Eq. (12.9-5), row vectors (t_1^i, t_2^i, t_3^i) $(i=1, 2, 3)$]. A vector $\boldsymbol{v}^{(0)}$ parallel to the 0-lines must result from the vector product of two of the row vectors

$$\boldsymbol{v}^{(0)} = [\boldsymbol{t}^{(i)} \times \boldsymbol{t}^{(k)}] \qquad (12.9\text{-}9)$$

(i) and (k) indicate the fixed indices. $\boldsymbol{t}^{(i)}$ and $\boldsymbol{t}^{(k)}$ are, as already mentioned, vectors in the reciprocal lattice 1 and $\boldsymbol{v}^{(0)}$ is a vector in the crystal coordinate system of lattice 1. The vector product procedure [Eq. (12.9-9)] may be verified by formulating the equations of the planes, e.g.:

$$\begin{aligned} t_1^1 v^1 + t_2^1 v^2 + t_3^1 v^3 &= 0 \\ t_1^2 v^1 + t_2^2 v^2 + t_3^2 v^3 &= 0\,. \end{aligned} \qquad (12.9\text{-}10)$$

It is then assumed that $v^3 = 1$, and v^1 and v^2 may be calculated. It turns out that the components are proportional to the vector product of the row vectors $t_k^{(i)}$.

The next task is to determine the normal to the b-subspace. For this purpose we consider Eq. (12.9-5) as a general relation between the x- and the b-space. For the case of rank $(\boldsymbol{T}) = 2$, the three-dimensional x-space is imaged onto the two-dimensional b-subspace, and consequently the image of every vector $\boldsymbol{x}$, and in particular the images of the $\boldsymbol{x}$-unit vectors $\boldsymbol{x} = \langle 100 \rangle$ must lie in the b-subspace. This means that the *column vectors* of $\boldsymbol{T}$ lie in the b-subspace and hence their vector product forms the normal vector $(c_1\, c_2\, c_3)$, the coordinates of which are specified in the reciprocal coordinate system.

$$\boldsymbol{c} = [\boldsymbol{t}_{(i)} \times \boldsymbol{t}_{(k)}]\,. \qquad (12.9\text{-}11)$$

Summarizing, we obtain for rank $(\boldsymbol{T}) = 2$ the orientation of the 0-lines (vector $\boldsymbol{v}^{(0)}$) from the vector product of two row vectors of the $\boldsymbol{T}$-matrix and the normal on the b-subspace, i.e. the coefficients of the equation for the b-subspace [Eq. (12.9-8)] from the vector product of two column vectors of the $\boldsymbol{T}$-matrix.

The vector $\boldsymbol{v}$ and $\boldsymbol{c}$ can be normalized to unit vectors:

$$|\boldsymbol{v}|^2 = (\boldsymbol{v}^T \boldsymbol{G} \boldsymbol{v}) \qquad (12.9\text{-}12)$$

$$|\boldsymbol{c}|^2 = (\boldsymbol{c}^T \boldsymbol{G}^{-1} \boldsymbol{c})\,. \qquad (12.9\text{-}13)$$

In order to determine the exact position of the 0-lines we require additional particular solutions of Eq. (12.9-5). We obtain these, for example, by setting $x^3 = 0$ and calculating $x^{1(P)}$ and $x^{2(P)}$ (P = particular) from the first two equations

$$
\begin{aligned}
x^{1(P)} &= \frac{b^{1(L)} t_2^2 - b^{2(L)} t_2^1}{t_1^1 t_2^2 - t_2^1 t_1^2} \\
x^{2(P)} &= \frac{b^{2(L)} t_1^1 - b^{1(L)} t_1^2}{t_1^1 t_2^2 - t_2^1 t_1^2} \\
x^{3(P)} &= 0 .
\end{aligned} \tag{12.9-14}
$$

The $b^{i(L)}$ have to be the coordinates of lattice points of the b-lattice and they have to lie in the b-subspace, i.e. they must satisfy Eq. (12.9-8). Then the O-lines become:

$$
\begin{aligned}
x^{1(0)} &= \lambda v^1 + x^{1(P)} \\
x^{2(0)} &= \lambda v^2 + x^{2(P)} \\
x^{3(0)} &= \lambda v^3 + x^{3(P)}
\end{aligned} \tag{12.9-15}
$$

with λ as a continuous parameter which determines the coordinates of a point on an O-line.

In the case of an O-line lattice (rank $(\boldsymbol{T}) = 2$) a special representation of the situation perpendicular to the O-lines will be of advantage. Here every O-line appears as a point, and the essential features involved in the choice of a boundary (such as line density, cell structure, etc.) become directly visible. We call this the *standard form.*

In order to find the standard form we start from the basic equation

$$\boldsymbol{T}\boldsymbol{x}^{(0)} = \boldsymbol{b}^{(L)} . \tag{12.9-16}$$

The O-line $\boldsymbol{v}^{(0)}$ [Eq. (12.9-9)] and the normal to the b-subspace $\boldsymbol{c}$ [Eq. (12.9-10)] are assumed to have been already determined and normalized. We now apply a rotation $\boldsymbol{R}$ such that $\boldsymbol{c}$ falls on the b_3-axis (the z-axis, respectively) [Appendix A1, Eq. (A1-63 ff.)]

$$\boldsymbol{R}\boldsymbol{T}\boldsymbol{x}^{(0)} = \boldsymbol{R}\boldsymbol{b}^{(L)} = \boldsymbol{b}'^{(L)} . \tag{12.9-17}$$

It is useful to formulate (12.9-16) in orthogonal coordinates since $\boldsymbol{R}$ then has the form (A1-60).

Now the b-subspace lies in the (orthogonal) $b_1 b_2$-plane. The O-lines were also rotated by this operation but they are still arbitrarily oriented in the x-space. They can even be parallel to the (xy)-plane [the $(b_1 b_2)$-plane respectively]. The orientation of the rotated O-lines is then given by:

$$\boldsymbol{R}\boldsymbol{v}^{(0)} = \boldsymbol{v}'^{(0)} . \tag{12.9-18}$$

Now we determine a second rotation $\boldsymbol{Q}$ which brings $\boldsymbol{v}'^{(0)}$ onto the z-axis

$$\boldsymbol{Q}\boldsymbol{v}'^{(0)} = \boldsymbol{v}''^{(0)} \tag{12.9-19}$$

i.e.

$$\boldsymbol{v}''^{(0)} = [0, 0, v_z''^{(0)}] .$$

Next we set

$$\boldsymbol{Q}\boldsymbol{x}^{(0)}=\boldsymbol{x}'^{(0)} \tag{12.9-20}$$

hence:

$$\boldsymbol{x}^{(0)}=\boldsymbol{Q}^{-1}\,\boldsymbol{x}'^{(0)}. \tag{12.9-21}$$

Along with Eq. (12.9-17), this gives

$$(\boldsymbol{RTQ}^{-1})\,\boldsymbol{x}'^{(0)}=\boldsymbol{b}'^{(L)}. \tag{12.9-22}$$

This equation states that every point in the $(b_1\,b_2)$-plane has as its image an x-line parallel to the z-axis. The representation of the 0-lines is independent of x_3. We can now reduce the system of equations from three to two dimensions by simply ignoring the third component.

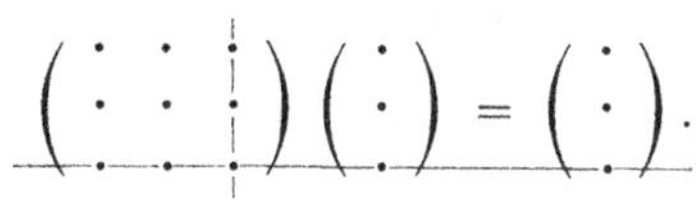

(12.9-23)

Thus, the matrix takes the form

$$(\boldsymbol{RTQ}^{-1})^i_k=t'^i_k \qquad (i,k=1,2) \tag{12.9-24}$$

and the Eq. (12.9-22) becomes

$$\begin{pmatrix} t'^1_1 & t'^1_2 \\ t'^2_1 & t'^2_2 \end{pmatrix}\begin{pmatrix} (x'^{(0)})^1 \\ (x'^{(0)})^2 \end{pmatrix}=\begin{pmatrix} (b'^{(L)})^1 \\ (b'^{(L)})^2 \end{pmatrix}. \tag{12.9-25}$$

The solutions are

$$\boldsymbol{x}'^{(0)}=\boldsymbol{T}'^{-1}\,\boldsymbol{b}'^{(L)}. \tag{12.9-26}$$

Here we have obtained the standard form for rank$(\boldsymbol{T})=2$ which represents the 0-lattice in the projection perpendicular to the 0-lines. The orientation of the 0-lines in the orthogonal, or in the crystal coordinate system is given by $\boldsymbol{v}^{(0)}$ [Eq. (12.9-9)].

In the case of rank$(\boldsymbol{T})=1$, the three planes represented by Eq. (12.9-5) are parallel. In order that solutions exist, the three planes have to coincide. Hence, only one b-component can be chosen freely, since the others are thereby fixed. This means that, in this case, the b-subspace is a straight line. For every b-lattice point on that line the solution of Eq. (12.9-5) is an *0-plane*. All three row vectors of $\boldsymbol{T}$ are parallel and, as such, are normal vectors of the 0-plane. On the other hand, the column vectors of $\boldsymbol{T}$ are also parallel and lie in the b-subspace, i.e. the b-line. Each of the three equations of (12.9-5) is an equation of the 0-plane if $\boldsymbol{b}^{(L)}$ is a lattice vector of the b-lattice and lies in the b-line, i.e. the b-subspace.

By a procedure analogous to that used for rank$(\boldsymbol{T})=2$ we can find a standard form for rank$(\boldsymbol{T})=1$. Here a normalized row vector which represents the normal to an 0-plane is denoted by $\boldsymbol{v}^{(0)}$, and a normalized column vector which determines the b-subspace by $\boldsymbol{c}$. We now rotate by an amount $\boldsymbol{R}$ the vector $\boldsymbol{c}$ onto the b_3-axis, and by an amount $\boldsymbol{Q}$ the vector $\boldsymbol{v}'^{(0)}$ onto the z-axis. We again obtain an equation of the type (12.9-22). In this case, however, it represents the relationship between the points on the b_3-axis and the z-position of the planes perpendicular to the z-axis, i.e. it represents a relationship between two one-dimensional sets of points. The equations are reduced to one dimension by

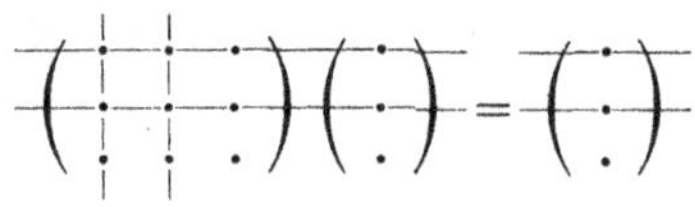

(12.9-27)

Hence:

$$t'^3_3 (x'^{(0)})^3 = (b'^{(L)})^3 \tag{12.9-28}$$

and the standard form for rank $(\boldsymbol{T}) = 1$ is:

$$(x'^{(0)})^3 = \frac{(b'^{(L)})^3}{t'^3_3}. \tag{12.9-29}$$

We designate the 0-points, 0-lines and 0-planes as 0-*elements*. An 0-element is the image of one b-lattice point.

The results of this section are presented in Table 12.11/T1. The 0-elements which are those regions where the two lattices show best fit, are points, lines, or planes depending on the rank $(\boldsymbol{T})$. 0-planes are, for example, twin planes. The (trivial) limiting case of rank $(\boldsymbol{T}) = 0$ signifies that the transformation $\boldsymbol{A}$ is equal to $\boldsymbol{I}$ i.e. we are dealing with a perfect crystal containing no interface.

In the case of rank $(\boldsymbol{T}) = 2$ we can interpret the vector $\boldsymbol{v}^{(0)}$, which fixes the orientation of the 0-line, as an invariant eigenvector of the transformation $\boldsymbol{A}$. An eigenvector maintains its orientation during the transformation but may change its length by a factor λ.

$$\boldsymbol{A}\boldsymbol{x} = \lambda \boldsymbol{x}. \tag{12.9-30}$$

If, in addition, it conserves its length we call it an *invariant eigenvector* ($\lambda = 1$), i.e.

$$\boldsymbol{A}\boldsymbol{x} = \boldsymbol{x}. \tag{12.9-31}$$

We write

$$\boldsymbol{A}\boldsymbol{x} = \boldsymbol{x}^{(0)} \tag{12.9-32}$$

(i.e. we refer to the end of the transformation $\boldsymbol{x}^{(0)}$ instead of to the initial $\boldsymbol{x}$). Thus:

$$\boldsymbol{x} = \boldsymbol{A}^{-1} \boldsymbol{x}^{(0)} \tag{12.9-33}$$

and, along with Eq. (12.9-31) this gives:

$$(\boldsymbol{I} - \boldsymbol{A}^{-1}) \boldsymbol{x}^{(0)} = 0 \tag{12.9-34}$$

which is the homogeneous part of the basic equation.

In the case of rank $(\boldsymbol{T}) = 1$ the transformation $\boldsymbol{A}$ contains two linearly independent invariant eigenvectors.

We have assumed in this section that be b-subspace contains b-lattice points. The problems which arise when this assumption is not valid will be discussed in Section 12.11.

12.10. The Behaviour of the 0-Lattice upon Translation of Lattice 2

If, in one of the moiré models, lattice 2 is translated with respect to lattice 1, the 0-lattice also experiences a translation. We shall now derive the analytical relations between the two translations.

We start again from the principle of the coincidence of elements of related equivalence classes. Lattice 2 shall be displaced by the vector $\boldsymbol{d}^{(2)}$.

$\boldsymbol{x}^{(1)}$ and $\boldsymbol{x}^{(1)} + \boldsymbol{b}^{(L)}$ are elements of the special class $C^{(1)}$. That element of the class $C^{(2)}$ which is related to $\boldsymbol{x}^{(1)}$ is no longer $\boldsymbol{A}\boldsymbol{x}^{(1)}$ but $\boldsymbol{A}\boldsymbol{x}^{(1)} + \boldsymbol{d}^{(2)}$. Let us imagine for the moment that $\boldsymbol{x}^{(1)}$ is a lattice point. Since lattice 2 is displaced by $\boldsymbol{d}^{(2)}$ there is no longer a lattice point at the position $\boldsymbol{A}\boldsymbol{x}^{(1)}$; it is displaced to the position $\boldsymbol{A}\boldsymbol{x}^{(1)} + \boldsymbol{d}^{(2)}$. Thus, the equation for the coincidence of equivalent points is

$$\boldsymbol{A}\boldsymbol{x}^{(1)} + \boldsymbol{d}^{(2)} = \boldsymbol{x}^{(1)} + \boldsymbol{b}^{(L)} = \boldsymbol{x}^{(0d)} \tag{12.10-1}$$

$\boldsymbol{x}^{(0d)}$ means "displaced $\boldsymbol{x}^{(0)}$". From this it follows that:

$$\boldsymbol{x}^{(1)} = \boldsymbol{A}^{-1}\left(\boldsymbol{x}^{(0d)} - \boldsymbol{d}^{(2)}\right) \tag{12.10-2}$$

and

$$\boldsymbol{x}^{(0d)} = \boldsymbol{A}^{-1}\boldsymbol{x}^{(0d)} - \boldsymbol{A}^{-1}\boldsymbol{d}^{(2)} + \boldsymbol{b}^{(L)} \tag{12.10-3}$$

$$(\boldsymbol{I} - \boldsymbol{A}^{-1})\,\boldsymbol{x}^{(0d)} = \boldsymbol{b}^{(d)} \tag{12.10-4}$$

with

$$\boldsymbol{b}^{(d)} = \boldsymbol{b}^{(L)} + \boldsymbol{d}^{(b)} = \boldsymbol{b}^{(L)} - \boldsymbol{A}^{-1}\boldsymbol{d}^{(2)} \tag{12.10-5}$$

$\boldsymbol{d}^{(b)}$ is the displacement of the b-lattice. Hence:

$$\boldsymbol{d}^{(b)} = -\boldsymbol{A}^{-1}\boldsymbol{d}^{(2)}. \tag{12.10-6}$$

If $\boldsymbol{x}^{(0d)}$ is decomposed as follows

$$\boldsymbol{x}^{(0d)} = \boldsymbol{x}^{(0)} + \boldsymbol{d}^{(0)} \tag{12.10-7}$$

then:

$$(\boldsymbol{I} - \boldsymbol{A}^{-1})\,\boldsymbol{x}^{(0)} + (\boldsymbol{I} - \boldsymbol{A}^{-1})\,\boldsymbol{d}^{(0)} = \boldsymbol{b}^{(L)} + \boldsymbol{d}^{(b)} \tag{12.10-8}$$

and hence:

$$(\boldsymbol{I} - \boldsymbol{A}^{-1})\,\boldsymbol{d}^{(0)} = \boldsymbol{d}^{(b)} \tag{12.10-9}$$

or

$$\boldsymbol{T}\boldsymbol{d}^{(0)} = \boldsymbol{d}^{(b)}. \tag{12.10-10}$$

This latter relation is valid in any case for rank$(\boldsymbol{T}) = 3$ and will have to be rechecked for the other cases (Section 12.11). In summary, if lattice 2 is translated by $\boldsymbol{d}^{(2)}$, then the b-lattice is displaced by $\boldsymbol{d}^{(b)}$, with

$$\boldsymbol{d}^{(b)} = -\boldsymbol{A}^{-1}\,\boldsymbol{d}^{(2)} \tag{12.10-11}$$

and the 0-lattice is displaced by $\boldsymbol{d}^{(0)}$, with

$$\boldsymbol{T}\boldsymbol{d}^{(0)} = \boldsymbol{d}^{(b)}. \tag{12.10-12}$$

Upon translation of lattice 2 the 0-lattice is also translated. The geometrical structure of the 0-lattice, however, is conserved [as long as rank$(\boldsymbol{T}) = 3$].

We can also describe the situation in another manner by attributing individual coordinate systems to every one of the four lattices. The coordinate system of lattice 1 (and the b-lattice, respectively) and of lattice 2 are related by the transformation $\boldsymbol{A}$, and those of the b- and of the 0-lattice are related by the transformation $\boldsymbol{T}$. If lattice 2 is translated by a vector $\boldsymbol{d}^{(2)}$ with the coordinates (d^1, d^2, d^3) in the coordinate system of lattice 2, then the b-lattice is translated by $\boldsymbol{d}^{(b)}$ with the coordinates $(-d^1, -d^2, -d^3)$ in the b-lattice coordinate system, and, correspondingly, the 0-lattice by $d^{(0)}$ $(-d^1, -d^2, -d^3)$ in terms of its own coordinate system. Apart from the sign, the numerical values of all the d-coordinates are the same but in different coordinate systems (Fig. 12.10/1). These relations can be checked with the aid of the moiré

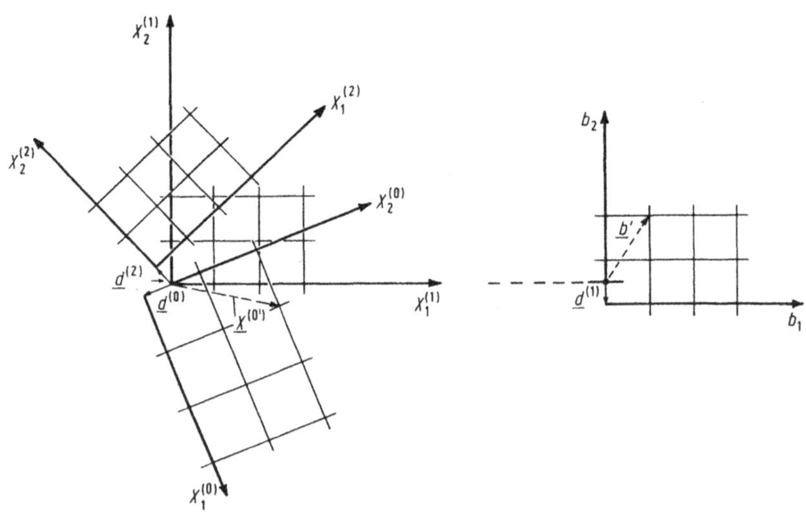

Fig. 12.10/1. 0-lattice translation resulting from the translation of lattice 2

models. By introducing the translation we have extended the range of the transformations $\boldsymbol{A}$ to nonhomogeneous transformations.

Special problems appear for translation in the case of rank$(\boldsymbol{T}) < 3$. These will be discussed in the following section.

12.11. Translation of the 0-Lattice for the Case of Rank $(T) < 3$

In Section 12.10 the relations between a translation of lattice 2 and the translation of the b- and the 0-lattices were derived. They are as follows:

$$\boldsymbol{d}^{(b)} = -\boldsymbol{A}^{-1}\,\boldsymbol{d}^{(2)} \tag{12.11-1}$$

and

$$\boldsymbol{T}\boldsymbol{d}^{(0)} = \boldsymbol{d}^{(b)}. \tag{12.11-2}$$

If rank$(\boldsymbol{T}) = 3$:

$$\boldsymbol{d}^{(0)} = \boldsymbol{T}^{-1}\,\boldsymbol{d}^{(b)}. \tag{12.11-3}$$

Thus, when the 0-lattice is a point lattice, it is uniquely determined.

Let us now consider the case of rank$(\boldsymbol{T}) = 2$, where only those b-lattice points which lie within the b-subspace (i.e. the b-plane) have 0-line images. Hence, we have to split the translation of the b-lattice $\boldsymbol{d}^{(b)}$ into two parts, one parallel and the other perpendicular to the b-plane.

$$\boldsymbol{d}^{(b)} = \boldsymbol{d}^{(b\parallel)} + \boldsymbol{d}^{(b\perp)}. \tag{12.11-4}$$

If $\boldsymbol{d}^{(b\perp)} = 0$, then $\boldsymbol{d}^{(b)}$ translates a b-lattice point within the b-plane by $\boldsymbol{d}^{(b\parallel)}$, and accordingly the corresponding 0-line is translated by $\boldsymbol{d}^{(0)}$. This means that the 0-line lattice is translated as a whole entity, and the problem may be treated as a two-dimensional one.

However, the action of $\boldsymbol{d}^{(b\perp)}$ is different. In the case where a certain b-lattice point was originally located within the b-plane and had an 0-line image, then, because of the displacement by $d^{(b\perp)}$ it leaves the b-plane and the corresponding 0-line disappears (Fig. 12.11/1).

As an example we discuss the rotation of a cubic lattice around the [001]-axis. That part of the b-lattice which lies in the b-subspace is the square point lattice in the (001)-plane. The interpenetrating lattices form layers of points in integral x_3-positions. The 0-lines are parallel to the x_3-axis. Every point on an 0-line is a coincidence of equivalent points in both lattices, and as such they represent possible origins for the transformation $\boldsymbol{A}$ which is here assumed to be the rotation (homogeneous transformation). If lattice 2 is now translated, for example, by $\boldsymbol{d}^{(2)} =$ [0, 0, 1/4], then the b-lattice is translated by [0, 0, −1/4]. The b-subspace ($b_3 = 0$) no longer contains b-lattice points. On the other hand, the two crystal lattices, where lattice 2 is now displaced with respect to

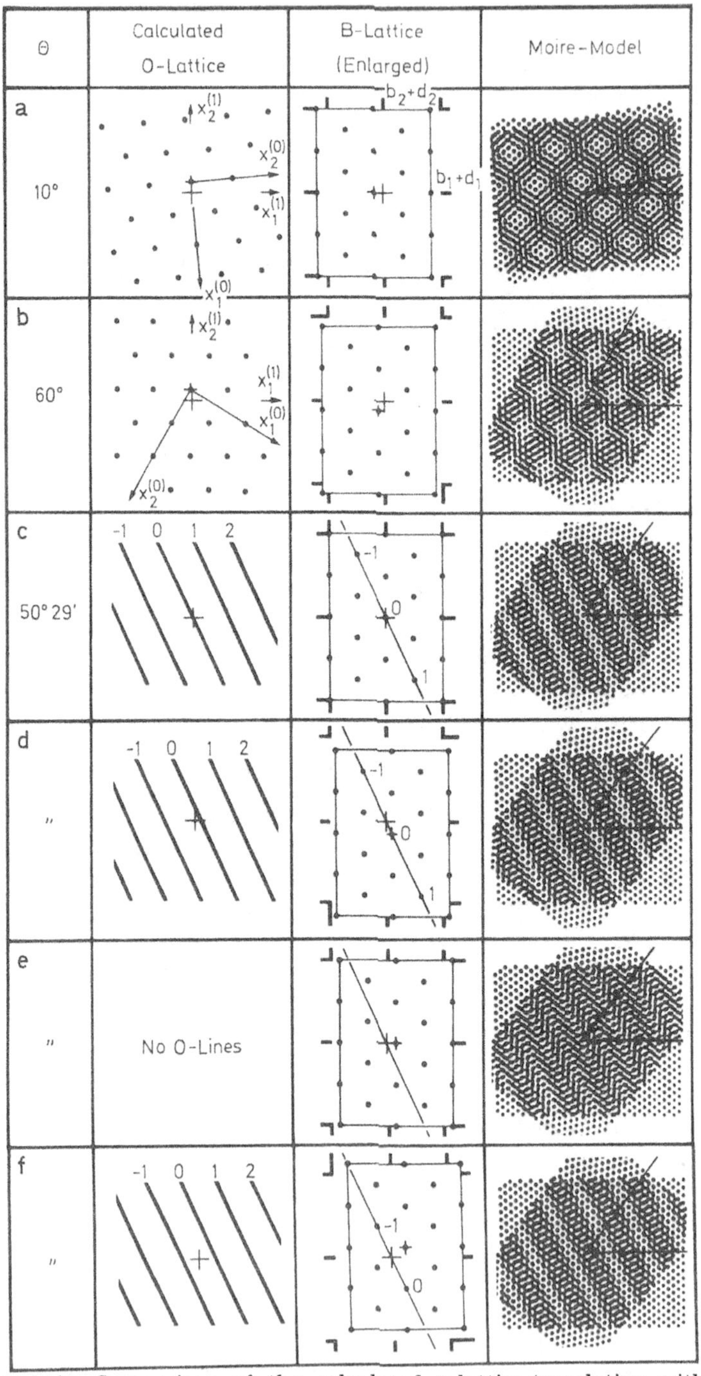

Fig. 12.11/1. Comparison of the calculated O-lattice-translation with the moiré model No. 1

lattice 1, cannot be imaged onto one another by the pure rotation, i.e. there are no coincidences of equivalent points i.e. no 0-lines.

An analogous situation appears in the case of rank$(\boldsymbol{T}) = 1$. Here the translation $\boldsymbol{d}^{(b)}$ of the b-lattice has to be split into one component parallel to the 0-line and another perpendicular to it. The translation of the b-lattices provokes also here a corresponding translation or disappearance of the 0-lattice.

Now we are in the position to solve the important problem which arises if the b-subspace contains few b-lattice points or none at all. In this case, the homogeneous transformation $\boldsymbol{A}$ could produce an 0-lattice consisting of only a few 0-elements. The 0-lattice would be "diluted". On the other hand, the b-vectors between the b-lattice points (i.e. the Burgers vectors of the dislocations between the 0-elements) would become very large. In reality, however, a "dense" 0-lattice will appear with elementary b-vectors. For this reason, 0-elements will also have to originate from b-lattice points in the close surroundings of the b-subspace. We project these b-lattice points perpendicularly onto the b-subspace, where the component $\boldsymbol{b}^{(L\parallel)}$ determines the position of the 0-element, and the component $\boldsymbol{b}^{(L\perp)}$ corresponds to a translation of lattice 2 with respect to lattice 1. The 0-lattice is determined here by a nonhomogeneous transformation where the homogeneous part is the same for all 0-elements, while the translation may be different for each 0-element.

We consider the case of rank$(\boldsymbol{T}) = 2$ and split $\boldsymbol{b}^{(L)}$ as follows

$$\boldsymbol{b}^{(L)} = \boldsymbol{b}^{(L\parallel)} + \boldsymbol{b}^{(L\perp)} \tag{12.11-5}$$

with

$$\boldsymbol{b}^{(L\parallel)} = \boldsymbol{b}^{(L)} - (\boldsymbol{c}\boldsymbol{b}^{(L)})\,\boldsymbol{c} \tag{12.11-6}$$

$\boldsymbol{c}$ is the unit vector normal to the b-plane ($=b$-subspace)

$$\boldsymbol{b}^{(L\perp)} = (\boldsymbol{c}\boldsymbol{b}^{(L)})\,\boldsymbol{c}. \tag{12.11-7}$$

Now the basic equation becomes

$$(\boldsymbol{I} - \boldsymbol{A}^{-1})\,\boldsymbol{x}^{(0)} = \boldsymbol{b}^{(L)} - (\boldsymbol{c}\boldsymbol{b})^{(L)}\boldsymbol{c} \tag{12.11-8}$$

and the translation

$$\boldsymbol{d}^{(b\perp)} = -\boldsymbol{b}^{(L\perp)} = -(\boldsymbol{c}\boldsymbol{b}^{(L)})\,\boldsymbol{c} = -\boldsymbol{A}^{-1}\boldsymbol{d}^{(2)}. \tag{12.11-9}$$

The change of sign: $\boldsymbol{d}^{(b\perp)} = -\boldsymbol{b}^{(L\perp)}$ arises because the b-lattice point is translated from outside into the b-subspace, which is considered to be fixed; hence, $\boldsymbol{d}^{(b\perp)}$ points towards the b-plane, while $\boldsymbol{b}^{(L\perp)}$ points away from the b-plane (Fig. 12.11/2a).

Thus:

$$\boldsymbol{d}^{(2)} = \boldsymbol{A}\,(\boldsymbol{c}\boldsymbol{b}^{(L)})\,\boldsymbol{c}. \tag{12.11-10}$$

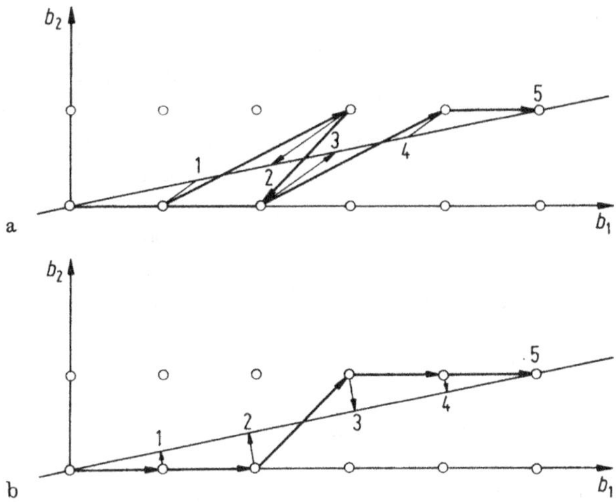

Fig. 12.11/2a and b. Projection of the b-lattice points onto the b-subspace. a) Perpendicular projection, b) projection along the 0-lines in a general case

It was proposed earlier (Bollmann, 1967, p. 376) that the b-lattice point be projected along the 0-line. Here, however, lattice 2 would also have been translated along that direction, thus eventually leading to physically unstable situations (Fig. 12.11/2b and Section 9.3) especially if the 0-lines were steeply inclined to $\boldsymbol{c}$. Therefore, it is advantageous to apply the shortest translation, i.e. the one perpendicular to the b-subspace, since the elastic energy is proportional to the square of this displacement. When $\boldsymbol{A}$ is a rotation, the two cases are identical.

For the case of rank$(\boldsymbol{T})=1$, the unit vector $\boldsymbol{s}$ shall point in the direction of the b-line (Section 12.9). Then:

$$\boldsymbol{b}^{(L\parallel)}=(\boldsymbol{s}\boldsymbol{b}^{(L)})\,\boldsymbol{s} \tag{12.11-11}$$

and

$$\boldsymbol{b}^{(L\perp)}=\boldsymbol{b}^{(L)}-(\boldsymbol{s}\boldsymbol{b}^{(L)})\,\boldsymbol{s}. \tag{12.11-12}$$

Hence

$$(\boldsymbol{I}-\boldsymbol{A}^{-1})\,\boldsymbol{x}^{(0)}=(\boldsymbol{s}\boldsymbol{b}^{(L)})\,\boldsymbol{s} \tag{12.11-13}$$

and

$$\boldsymbol{d}^{(2)}=\boldsymbol{A}[\boldsymbol{b}^{(L)}-(\boldsymbol{s}\boldsymbol{b}^{(L)})\,\boldsymbol{s}]. \tag{12.11-14}$$

By this procedure we have supported the stepped b-net of Section 11.6. In the general case, the b-plane (b-line resp.) in the b-lattice will be approximated by a stepped surface (line resp.) and the 0-lattice will be calculated by Eq. (12.11-8) [(12.11-13) resp.], the standard form may be obtained by applying (12.9-25) [(12.9-28) resp.].

Table 12.11/T1 summarizes the procedures for the solution of the basic equation.

Table 12.11/T 1

Rank ($\boldsymbol{T}$)	0-Elements (nD = n-dimensional)	Determination of the 0-Elements	b-Subspace	Determination of the b-Subspace	Determination of 0-Elements if $\boldsymbol{b}^{(L)}$ does not lie in the b-Subspace
3	0-point (0D)	$\boldsymbol{x}^{(0)} = \boldsymbol{T}^{-1}\,\boldsymbol{b}^{(L)}$	b-space (3D)	$\boldsymbol{b}^{(L)}$ arbitrary	
2	0-line (1D)	Vector product of row vectors of $\boldsymbol{T}$ + particular solution	b-plane (2D)	Normal to b-plane = vector product of column vectors of $\boldsymbol{T}$	Normal projection of $\boldsymbol{b}^{(L)}$ onto the b-plane
1	0-plane (2D)	Normal to 0-plane = row vectors of $\boldsymbol{T}$ + particular solution	b-line (1D)	Column vector of $\boldsymbol{T} = b$-line	Normal projection of $\boldsymbol{b}^{(L)}$ on b-line
0	0-space (3D) (perfect crystal)	Total x-space	b-point (0D)	Point $b\,[0, 0, 0]$	Translation into b-point (lattices 1 and 2 coincide perfectly)

12.12. Subdivision of the Crystal Space into Cells

The transformation $\boldsymbol{A}$ only relates the nearest neighbours in the "region" of an 0-element. The different 0-elements regions can be separated by cell walls. If the unit cell of the 0-lattice is large compared to that of the crystal lattice, these cell walls have a physical meaning. The lines of intersection of the grain- or phase boundary with the cell walls are the dislocations. Hence, the cell walls have to be carefully chosen.

We introduce the following criterion for the cell wall: We consider two 0-points on adjacent 0-elements as origins, particularly the point [000] and the point $\boldsymbol{x}^{(0)}$. The general vector $\boldsymbol{b}$ was defined by Eq. (11.1-2) as:

$$\boldsymbol{b} = \boldsymbol{x}^{(2)} - \boldsymbol{x}^{(1)} \tag{12.12-1}$$

i.e. as the difference vector of corresponding points in the two lattices. $\boldsymbol{x}^{(2)}$ is regarded here (somewhat arbitrarily!) as variable $\boldsymbol{x}$, i.e.

$$\boldsymbol{b}(\boldsymbol{x}) = \boldsymbol{x} - \boldsymbol{x}^{(1)}. \tag{12.12-2}$$

Now we compare the absolute values of these $\boldsymbol{b}$-vectors, on choosing the two 0-points as origins. $\boldsymbol{x}$ determines a point on the cell wall if

$$|\boldsymbol{b}(\boldsymbol{x})|^2 = |\boldsymbol{b}(\boldsymbol{x} - \boldsymbol{x}^{(0)})|^2. \tag{12.12-3}$$

The imaging relation between the b-space and the crystal space is given by

$$\boldsymbol{b}(\boldsymbol{x}) = (\boldsymbol{I} - \boldsymbol{A}^{-1})\,\boldsymbol{x} := \boldsymbol{T}\boldsymbol{x} \tag{12.12-4}$$

and

$$\boldsymbol{b}(\boldsymbol{x}^{(0)}) = \boldsymbol{b}^{(L)} = \boldsymbol{T}\boldsymbol{x}^{(0)}. \tag{12.12-5}$$

Under the condition defined by (12.12-4), we can interpret Eq. (12.12-3) as an imaging of Wigner-Seitz cells (Section 2.2) of the b-lattice onto the crystal lattice.

In arbitrary (crystal) coordinates:

$$|\boldsymbol{b}|^2 = \boldsymbol{b}^T \boldsymbol{G} \boldsymbol{b} \tag{12.12-6}$$

$\boldsymbol{b}^T$ = transposed vector of $\boldsymbol{b}$ (line vector),

$\boldsymbol{G}$ = metric tensor.

The condition (12.12-3) becomes:

$$|\boldsymbol{b}|^2 - |\boldsymbol{b} - \boldsymbol{b}^{(L)}|^2 = 0. \tag{12.12-7}$$

In matrix form:

$$\boldsymbol{b}^T \boldsymbol{G} \boldsymbol{b} - (\boldsymbol{b} - \boldsymbol{b}^{(L)})^T \boldsymbol{G} (\boldsymbol{b} - \boldsymbol{b}^{(L)}) = 0 \tag{12.12-8}$$

with

$$(\boldsymbol{b}-\boldsymbol{b}^{(L)})^T=\boldsymbol{b}^T-\boldsymbol{b}^{(L)T}. \tag{12.12-9}$$

From (12.12-8) follows that

$$\boldsymbol{b}^T\boldsymbol{G}\boldsymbol{b}-\boldsymbol{b}^T\boldsymbol{G}\boldsymbol{b}+\boldsymbol{b}^{(L)T}\boldsymbol{G}\boldsymbol{b}+\boldsymbol{b}^T\boldsymbol{G}\boldsymbol{b}^{(L)}-\boldsymbol{b}^{(L)T}\boldsymbol{G}\boldsymbol{b}^{(L)}=0. \tag{12.12-10}$$

The first two terms cancel one another. Since all the terms are scalars they are equal to their transpose. Because $(\boldsymbol{A}\boldsymbol{B})^T=\boldsymbol{B}^T\boldsymbol{A}^T$:

$$\boldsymbol{b}^T\boldsymbol{G}\boldsymbol{b}^{(L)}=[\boldsymbol{b}^T\boldsymbol{G}\boldsymbol{b}^{(L)}]^T=\boldsymbol{b}^{(L)T}\boldsymbol{G}^T\boldsymbol{b}=\boldsymbol{b}^{(L)T}\boldsymbol{G}\boldsymbol{b}. \tag{12.12-11}$$

The last step is possible because $\boldsymbol{G}$ is always symmetrical and consequently $\boldsymbol{G}^T=\boldsymbol{G}$. Now Eq. (12.12-10) becomes:

$$\boldsymbol{b}^{(L)T}\boldsymbol{G}\boldsymbol{b}-(1/2)\,\boldsymbol{b}^{(L)T}\,\boldsymbol{G}\boldsymbol{b}^{(L)}=0 \tag{12.12-12}$$

or, using Eq. (12.12-4):

$$\underline{(\boldsymbol{b}^{(L)T}\boldsymbol{G}\boldsymbol{T})\,\boldsymbol{x}-(1/2)\,(\boldsymbol{b}^{(L)T}\boldsymbol{G}\boldsymbol{b}^{(L)})=0.} \tag{12.12-13}$$

This is the general equation of a cell wall between two 0-elements. The cell wall bisects (but does not have to be perpendicular to) the connection between the 0-elements [$\boldsymbol{x}=(1/2)\ \boldsymbol{x}^{(0)}$ fulfills the equation]. [See also Eqs. (12.8-25) to (12.8-30).]

Fig. 13.3/3a presents an example of cell walls calculated for the case of a $\{1\,1\,0\}$-plane in the bcc structure for a rotation of 38° 56.6′. We have chosen $\boldsymbol{x}^{(2)}$ as variable $\boldsymbol{x}$, but we could equally have chosen $\boldsymbol{x}^{(1)}$ or some point between $\boldsymbol{x}^{(2)}$ and $\boldsymbol{x}^{(1)}$. In general, the orientation of the cell wall depends on that choice.

Chapter 12. Summary and Discussion

The geometrical theory of general crystalline interfaces consists essentially of two parts. The first part is the determination of the 0-lattice, its behaviour on translation of one crystal with respect to the other, and the calculation of the cell structure. The structure of the 0-lattice is located within the two interpenetrating translation lattices.

The second part of the theory concerns the problems connected with the choice of the crystal boundary. The boundary is placed in the 0-lattice, passing as much as possible through 0-elements, i.e. through the locations of best fit. Then crystal 1 is occupied by atoms on one side of the boundary and crystal 2 by atoms on the other side. Only at the boundary itself do some remnants of the 0-lattice structure exist. The 0-lattice is a general three-dimensional description of all the *possible* two-dimensional boundaries.

In this section, only the first part, i.e. the construction of the 0-lattice, is summarized. The essential steps are as follows:

1. Both crystal lattices are related by a linear transformation such that every lattice point of lattice 2 is the image of a lattice point of lattice 1 (Sections 12.2 and 12.3)

$$\boldsymbol{x}^{(2L)} = \boldsymbol{A}\boldsymbol{x}^{(1L)}.$$

2. The 0-elements are given as solutions of the basic equation

$$(\boldsymbol{I} - \boldsymbol{A}^{-1})\,\boldsymbol{x}^{(0)} = \boldsymbol{b}^{(L)}$$

where the $\boldsymbol{b}^{(L)}$-vectors are translation vectors of lattice 1, and as such are lattice vectors of the b-lattice (Section 12.3—12.5).

3. In the proximity of the origin, the transformation $\boldsymbol{A}$ must relate the *nearest neighbours* in the two lattices. If there is a choice between several possible transformations then that transformation which gives the smallest absolute value of the determinant $|\boldsymbol{I} - \boldsymbol{A}^{-1}|$ is the one which has the greatest physical significance. The formulation of the transformation $\boldsymbol{A}$ is given in Sections 12.7 and 12.8.

4. Depending on the rank of the matrix $(\boldsymbol{I} - \boldsymbol{A}^{-1}) := \boldsymbol{T}$, the solutions for $\boldsymbol{x}^{(0)}$ may be *points* [rank$(\boldsymbol{T}) = 3$], *lines* [rank$(\boldsymbol{T}) = 2$], or *planes* [rank$(\boldsymbol{T}) = 1$]. For rank$(\boldsymbol{T}) = 3$ images exist for all the lattice points of the b-lattice, while for rank$(\boldsymbol{T}) = 2$ and rank$(\boldsymbol{T}) = 1$ only those points within a b-plane and within a b-line, respectively, are imaged (Section 12.9).

5. A translation of lattice 2 by $\boldsymbol{d}^{(2)}$ in terms of the coordinates of lattice 2 involves a translation of the b-lattice by $-\boldsymbol{d}^{(b)}$, where $-\boldsymbol{d}^{(b)}$ has the same coordinate values in terms of the coordinates of the b-lattice (apart from the sign) as $\boldsymbol{d}^{(2)}$ in lattice 2. $-\boldsymbol{d}^{(b)}$ induces a displacement of the 0-lattice by $-\boldsymbol{d}^{(0)}$, the coordinates of which, in the coordinate system of the 0-lattice, are again numerically identical to the others in their respective coordinate system (Section 12.10).

6. If the b-subspace is only sparsely occupied by b-lattice points, it can be approximated by a stepped b-net. This also includes the b-lattice points in the close proximity of the b-subspace. This means that, for the transformation $\boldsymbol{A}$, a translation is added to the homogeneous transformation (Section 12.11).

7. In order to determine the cell walls between the 0-elements, it was stipulated that, at the cell wall

$$|\boldsymbol{b}(\boldsymbol{x})|^2 = |\boldsymbol{b}(\boldsymbol{x} - \boldsymbol{x}^{(0)})|^2$$

which leads to Eq. (12.12-13) and is equivalent to an imaging of Wigner-Seitz walls of the b-lattice onto the 0-lattice (Section 12.12).

The equation for the cell walls is plausible but not completely sure. Since the intersections of the interface with the cell walls represent the dislocations (as was already shown in Section 11.2) it is possible that elastic anisotropy has to be considered. Here we have chosen $\boldsymbol{x}^{(2)}$ as variable $\boldsymbol{x}$. Once the choice of $\boldsymbol{x}^{(1)}$, or a point between $\boldsymbol{x}^{(2)}$ and $\boldsymbol{x}^{(1)}$ is made, the orientation of the cell wall may vary, but it must always bisect neighbouring 0-elements.

For given structures and orientations of the two crystal lattices, all the elements of the 0-lattice are determined with the aid of the indications given in this chapter. The applications are dealt with in the in following chapter.

Suggestions for Problems

There is no lack of possible problems in this chapter. The calculations of Sections 12.7 and 12.8 can be carried out explicitly on given examples, such as moiré models Nos. 1, 2, and 3. The transformations which relate nearest neighbours can be determined. The calculated 0-lattices can be quantitatively compared with the moiré models. Furthermore, the examples of rotation of moiré model No. 1 ({1 1 0}-plane, bcc structure) by 50° 29′ and 70° 32′ can be reduced to the standard form. Similar possibilities exist with moiré model No. 2, for $\theta = 34°\,03'$ and 62° 58′. (The necessary indications are to be found in Table 13.3/T1.)

In addition, with respect to the translation and the cell structure, a sufficient number of examples can be found in connection with the moiré models (Sections 12.10 to 12.12): From the behaviour upon translation by $\theta = 60°$ of moiré model No. 1 ({1 1 0} bcc), for example, it can be shown that the coordinate system of the 0-lattice is left-handed ($|\boldsymbol{T}| < 0$).

However, care must be taken when using the moiré models, since translation operations carried out by means of sheet $\boldsymbol{a}$ (lattice 1) correspond to $\boldsymbol{d}^{(1)} = \boldsymbol{d}^{(b)}$ instead of $\boldsymbol{d}^{(2)}$ ($\boldsymbol{d}^{(1)} = -\boldsymbol{A}^{-1}\,\boldsymbol{d}^{(2)}$). Rotation, on the other hand, is done with sheet $\boldsymbol{b}$ (lattice 2).

13. Applications of the 0-Lattice Theory

13.1. Outline

Up till now the two adjoining crystals have been idealized as two interpenetrating point lattices. The 0-lattice with its cell structure introduces an ordering into the interpenetrating lattices. The next step is the choice of the boundary. Once the boundary is chosen, the atoms will occupy, on the one side of the boundary the positions of crystal 1, and on the other side those of crystal 2. The boundary itself is a remnant of the 0-lattice structure.

It shall be emphasized once again that the 0-lattice is a description of those positions where the two crystals fit in an optimum manner. The cell walls, on the other hand, are the positions of worst fit, i.e. the positions where the correlation between nearest lattice points in the two lattices changes between two 0-elements (origins!).

A crystal boundary will be placed, as far as possible, through 0-elements. However, not all the possible boundaries are equivalent, since the energy might vary. The problems related to the boundary energy will have to be discussed.

Often the problem is not only to find some boundary for some given relative orientation of the two crystals, but also to first determine the crystal orientation such that the best possible boundary can be drawn. Here, neither the relative orientation of the crystals nor that of the boundary is known at the beginning. We require a criterion for the optimum boundary, which is basically the criterion for the minimum boundary energy. In most cases it is not possible to calculate accurately enough the boundary energy. It will be shown that in certain cases an energy minimum can be determined without knowledge of the value of the energy.

We distinguish roughly two ranges of problems according to the dimensions of the 0-lattice unit cell compared to those of the crystal lattice.

1. The unit cell of the 0-lattice is large compared to that of the crystal lattice, and

2. the 0-lattice unit cell and the crystal lattice unit cell are of similar dimensions.

In the first case the two crystals deviate only slightly from one another. The crystal boundaries are formed by dislocation networks where the intersections of the boundary with the 0-lattice cell walls (Section 12.12) are the dislocation lines. On choosing the crystal boundary, a surface, which may eventually be stepped, is traced in the b-lattice; this surface contains all those b-lattice points which correspond to the 0-elements intersected by the boundary. These selected b-lattice points

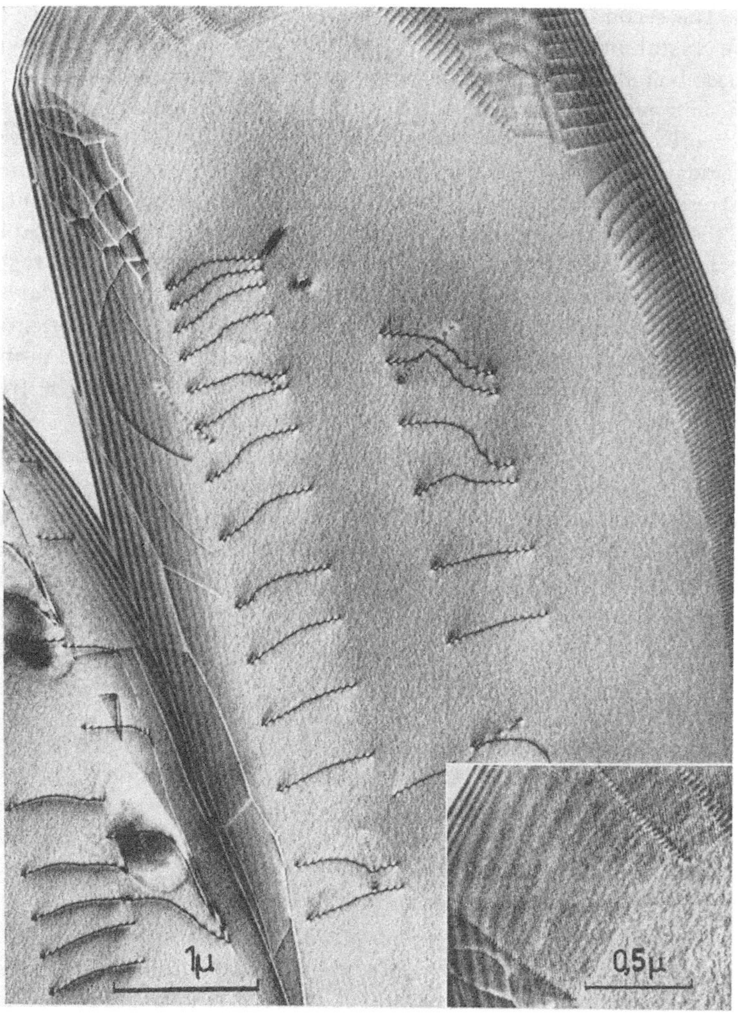

Fig. 13.1/1. Features in a grain boundary (stainless steel) as expected from the 0-lattice theory. (Photo: G. Dupouy, F. Perrier, Toulouse)

form the *b-net* of the dislocation network in the boundary. The Burgers vectors are attributed to the dislocation lines by the duality relations of Sections 9.1 and 11.2.

No optimum boundaries exist for subgrain boundaries, i.e. low angle boundaries between identical crystals. Here the optimum state is the perfect crystal, since the minimum energy of a subgrain boundary is at $\theta = 0$ (Section 11.3). However, optimum dislocation networks are possible in phase boundaries. An example will be discussed in Section 13.2.

The second example of when the dimensions of the O-lattice- and the crystal unit cells are similar, is the case of high angle- or the general phase-boundaries. Here the conventional notion of the dislocation loses its meaning since, if such a boundary is considered as a dislocation network, the dislocation cores would overlap. Nevertheless, optimum boundaries are also possible, but the emphasis has to be placed on the pattern of the lattice points: whether this pattern is periodic or not, and in the former case whether the period is large or small. It is to be expected that, for a small period, small groups of atoms come together and thus give rise to an energy minimum in the sense that any deviation from that boundary orientation, as well as from the relative orientation of the crystals, causes an increase in the boundary energy. The problems arising in connection with the pattern of lattice points will be treated in Section 13.3.

Fig. 13.1/1 shows an electron micrograph of a grain boundary in stainless steel taken with the 1.5 MeV microscope at Toulouse. The boundary shows the type of features as are excepted from the O-lattice theory. Further electron microscope observations have been reported and discussed by: Gleiter (1969), Ishida *et al.* (1969) and Levy (1969). Calculations on grain boundary energy were done by Fletcher (1967).

13.2. Example of an Optimum Phase Boundary

In this section we present an application of the O-lattice theory (Bollmann and Nissen, 1968), concerning the solution of a special mineralogical problem which, as such, should belong in a mineralogy text book. However, the exercise was carried out in order to test the O-lattice theory, because the situation was very general without any evident simplifications, and because measured data were available.

The problem concerns an exsolved alkali feldspar (Fig. 13.2/1), a lamellar two-phase system consisting of a monoclinic and a triclinic phase. The triclinic phase is twinned, which is evident from the vertical banding of that phase. The thickness of the lamellae is about 1,000 Å. The mineral acts as a natural optical grating which produces irridescent

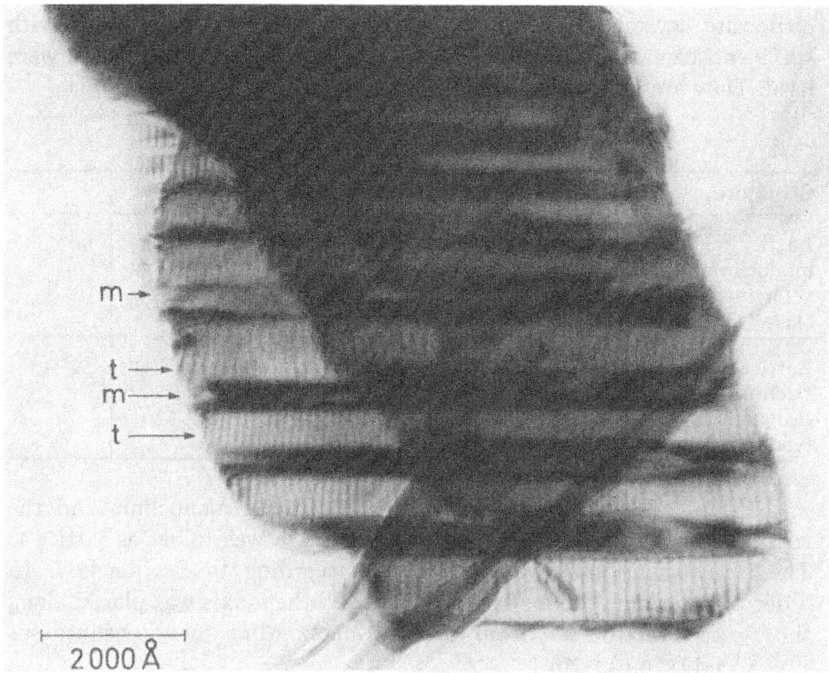

Fig. 13.2/1. Cryptoperthite (feldspar, moonstone) *m* monoclinic (homogeneous) *t* triclinic (twinned)

colours. For this reason it is often used as a gem stone under the name "moonstone".

Chemically speaking, this moonstone is essentially (K, Na) (Al, Si)$_3$O$_8$ and its structure is given in Fig. 13.2/2. Because of the fact that in the

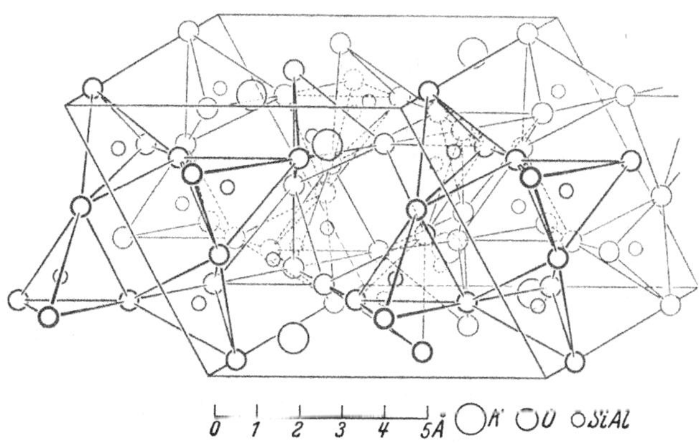

Fig. 13.2/2. The structure of feldspar

composite crystal the lattice constants could not be determined with sufficient accuracy, the values of the two most likely pure phases were used. They are given in Table 13.2/T1.

Table 13.2/T1

Structure	a (Å)	b (Å)	c(Å)	α	β	γ
Lattice 1 monoclinic ("Orthoclase" Or 92)	8.561(6)	12.996(2)	7.193(4)	90°	116° 1′	90°
Lattice 2 triclinic (Low albite Kodarma)	8.135(3)	12.788(3)	7.154(2)	94° 14′	116° 31′	87° 43′

First of all, an attempt was made to fit the monoclinic and the triclinic phases together. The monoclinic phase was taken as lattice 1. The transformation $\boldsymbol{A}$ was formulated according to Section 12.7. In order to determine the matrices $\mathbf{S}^{(1)}$ and $\mathbf{S}^{(2)}$ the a-axis was placed along the x-axis and the c-axis in the (xz)-plane. (For the procedure see Eqs. (A1-17) to (A1-26).]

For the determination of the optimum boundary, a rotation $\boldsymbol{R}$ of only a few degrees is required, since the two phases are very similar. The complete rotation can be expressed by three separate rotations around the x-, y- and z-axes. As long as the rotations are small enough, they can be superposed as polar vectors, i.e.

$$\boldsymbol{R} = \boldsymbol{R}_x \cdot \boldsymbol{R}_y \cdot \boldsymbol{R}_z . \tag{13.2-1}$$

Thus, in terms of orthogonal coordinates:

$$\boldsymbol{A} = \boldsymbol{R}_z \boldsymbol{R}_y \boldsymbol{R}_x \mathbf{S}^{(2)} \mathbf{S}^{(1)-1} . \tag{13.2-2}$$

First the determinant $|\boldsymbol{I} - \boldsymbol{A}^{-1}| = |\boldsymbol{T}|$ was calculated as a function of the angles ϕ_x, ϕ_y, ϕ_z, and it proved to be indefinite (Fig. 13.2/3). In the space with the angles of rotation as coordinates the surface $|\boldsymbol{T}| = 0$ resembles a single shell hyperboloid. Inside the hyperboloid $|\boldsymbol{T}| > 0$, and outside $|\boldsymbol{T}| < 0$.

No optimum is apparent in this representation. For all the orientations ϕ_i on the surface $|\boldsymbol{T}| = 0$, the 0-lattice is a line lattice, but also here the line density can strongly vary.

We have to remember that the determinant $|\boldsymbol{T}|$ is a measure of the volume of the 0-lattice unit cell, while what we require is a surface measure in order to be able to determine the optimum boundary. As has

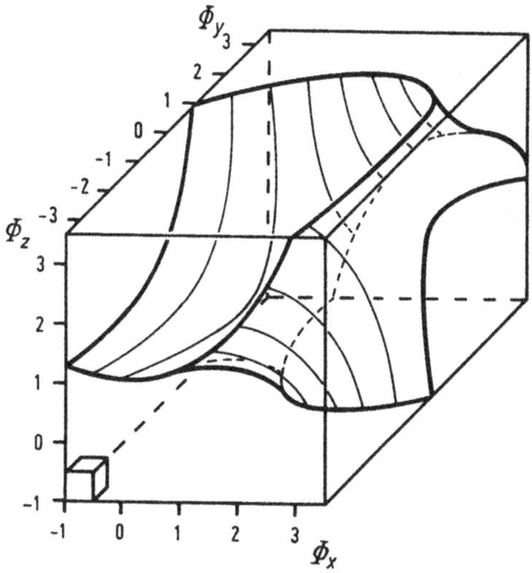

Fig. 13.2/3. Adaptation: monoclinic-triclinic $|\boldsymbol{I} - \boldsymbol{A}^{-1}| = 0$ as a function of the rotation ϕ of lattice 2 (units: degrees)

already been mentioned, the energy is not known. For energy calculations to be carried out, values of material constants would be needed, the accuracy of which would be doubtful. It is preferable to adhere strictly to the geometrical approach. From the considerations of subgrain boundaries, we know that the energy first increases with increasing dislocation density (Section 11.7). Therefore, we replace the energy function by a purely geometrical parameter, which we can assume to vary monotonically with the energy in the range of interest. It does not necessarily have to be proportional to the energy. We only require that it increases when the energy increases, and vice versa.

For a series of parallel dislocations of spacing d and with an absolute value of the Burgers vector b, we replace the energy by the parameter

$$P = \left(\frac{b}{d}\right)^2. \tag{13.2-3}$$

If the boundary is formed by a dislocation network consisting of two independent series of dislocations, then:

$$P = \sum_{i=1}^{2} \left(\frac{b_i}{d_i}\right)^2. \tag{13.2-4}$$

This function has a positive definite value, and remains continuous even when one of the dislocation series disappears, i.e. when one of the d_i becomes infinite.

This geometrical parameter is determined in the following manner: In the case of rank$(\boldsymbol{T}) = 3$ the 0-lattice is a point lattice. We assume as possible boundaries the three faces of the 0-lattice unit cell (parallelepiped). These boundaries pass through the 0-points, and their intersections with the cell walls are the dislocations. In order to simplify the problem, we assume the cell walls to be parallel to the faces of the 0-lattice unit cell (Fig. 13.2/4). The geometrical parameter P_1 is specified as:

$$P_1 = \frac{|\boldsymbol{b}_2|^2}{(d_2)^2} + \frac{|\boldsymbol{b}_3|^2}{(d_3)^2}. \tag{13.2-5}$$

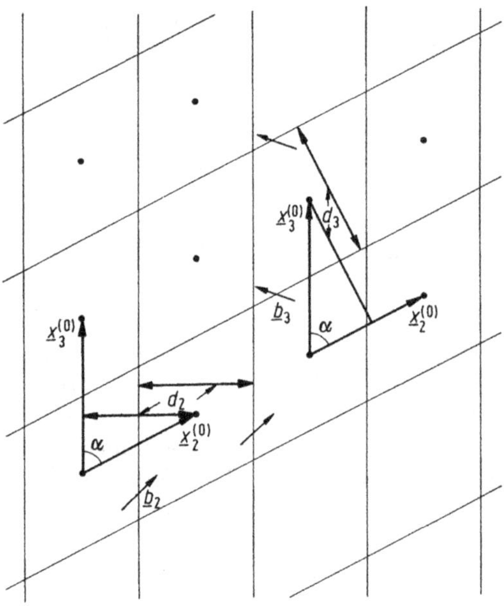

Fig. 13.2/4. Elements required for the determination of the geometrical parameter [Eq. (13.2-8)]

P_2 and P_3 follow from the cyclic permutation of the indices. d_2 is given by:

$$d_2 = \left| \boldsymbol{x}_2^{(0)} \times \frac{\boldsymbol{x}_3^{(0)}}{|\boldsymbol{x}_3^{(0)}|} \right| = \frac{1}{\boldsymbol{x}_3^{(0)}} |\boldsymbol{x}_2^{(0)} \times \boldsymbol{x}_3^{(0)}| \tag{13.2-6}$$

and:

$$\frac{|\boldsymbol{b}_2^{(L)}|^2}{(d_2)^2} = \frac{|\boldsymbol{b}_2^{(L)}|^2 \cdot |\boldsymbol{x}_3^{(0)}|^2}{|\boldsymbol{x}_2^{(0)} \times \boldsymbol{x}_3^{(0)}|^2} \tag{13.2-7}$$

and together with Eq. (13.2-5):

$$P_1 = \frac{|\boldsymbol{b}_2^{(L)}|^2 |\boldsymbol{x}_3^{(0)}|^2 + |\boldsymbol{b}_3^{(L)}|^2 |\boldsymbol{x}_2^{(0)}|^2}{|\boldsymbol{x}_2^{(0)} \times \boldsymbol{x}_3^{(0)}|^2} \tag{13.2-8}$$

and, correspondingly P_2 and P_3.

The orientation of the crystal lattices (ϕ_x ϕ_y ϕ_z), for which the smallest of the above parameters exhibited a minimum, was understood as being that orientation for which the 0-lattice furnishes the optimum boundary. The orientation of the boundary was given by the two $\boldsymbol{x}_i^{(0)}$-vectors which were contained in the expression defining that parameter. P_1 proved to be the smallest of the parameters, and its value is given as a function of the angle ϕ_i in Fig. 13.2/5. The surprising result is that this parameter

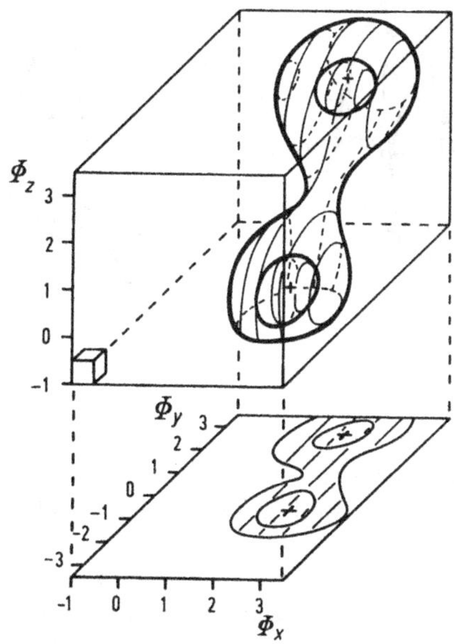

Fig. 13.2/5. Geometrical parameter $P1$ for the monoclinic-triclinic adaptation (coordinates as in Fig. 13.2/3)

exhibits two minima. The value of P_1 at the minima is about 26 units. The function is continuous on the surface $|\boldsymbol{T}| = 0$, and the minima lie very close to that surface, i.e. the 0-lattice of the optimum boundary is practically a line lattice. The orientations of the crystals and of the boundary determined above are completely different from the measured values.

It is to be noted that the triclinic phase is twinned. According to F. Laves this twinning is an averaging process which takes place in the triclinic phase in order to achieve a pseudo-monoclinic character. The twinning of the triclinic phase can be explained by one of two different twin laws, the "Albite law" (twin axis $\boldsymbol{b}^*$) and the "Pericline law" (twin axis $\boldsymbol{b}$). (The details are found in the indicated article.) Accordingly the matrix $\boldsymbol{S}^{(2)}$ was averaged, so that now two monoclinic phases had to be fitted together.

Fig. 13.2/6 shows the configuration corresponding to Figs. 13.2/3 and 13.2/5, for the fitting of the two monoclinic phases. Here, the determinant $|\boldsymbol{T}|$ has a negative definite value and its absolute value shows a distinct minimum, as in the case of P_1. Both minima coincide in the

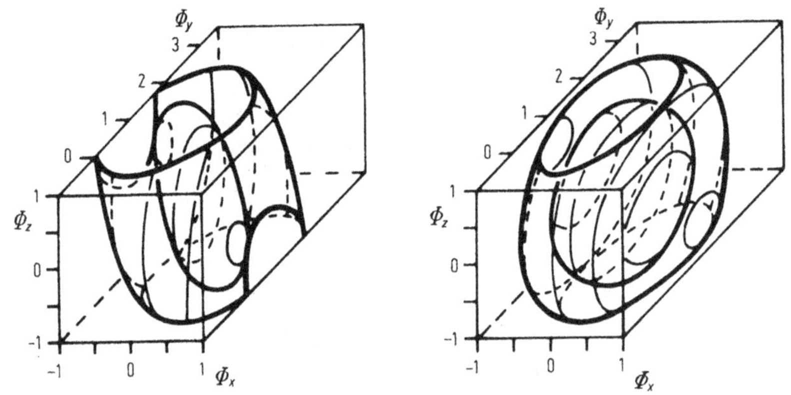

Fig. 13.2/6. a) $|\boldsymbol{I}-\boldsymbol{A}^{-1}|$ and b) $P1$ for the monoclinic-monoclinic adaptation

ϕ-space and, in addition, the value of P_1 is 3 units, i.e. about eight times smaller than in the case of Fig. 13.2/3 which indicates that here also the boundary energy is markedly smaller than in the first case. Fig. 13.7 shows the calculated optimum boundary. The calculated dislocation network might well represent the period in the boundary, but not the details. The dislocations will split into partials since a Burgers vector of 13 Å will probably not be stable. In addition, the boundary is disturbed by the twinning but is nevertheless stable. The comparison between the calculated and the measured values shows good agreement (Fig. 13.2/8). The differences in the averaging process carried out according to the two twin laws do not appear to be significant. However, in order to reduce the twin boundary surface to a minimum, the Albite twinning law is preferred in this case.

The calculation brings to light another property of the optimum boundary. The optimum transformation $\boldsymbol{A}(\phi_i^{(\mathrm{opt})})$ has three eigenvectors with the corresponding eigenvalues λ_i. The optimum boundary contains those two eigenvectors, with the eigenvalues λ_i, which show the smallest deviation from 1. In the boundary, the crystal experiences no rotation only extension, and this extension is minimal.

The result of the calculation is surprising if we consider under what assumptions it was obtained. The only numerical data used was the two sets of six lattice constants each of the two crystals (Table 13.2/T1); no

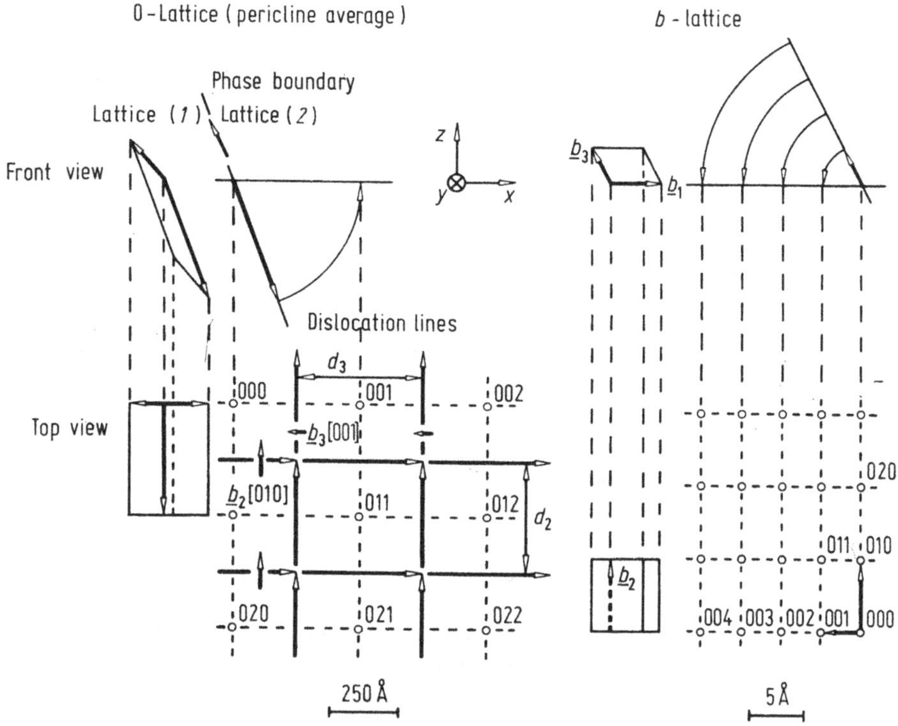

Fig. 13.2/7. O-lattice of the optimum boundary

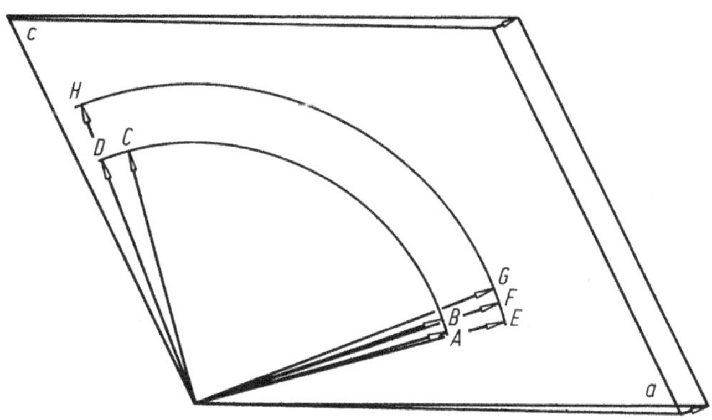

Fig. 13.2/8 A–H. Comparison between the calculated and the measured data. Relative positions of the unit cells, measured rotation: 55–60′; calculated: 56′. A: measured eigenvector (EV) [3, 0, 1]; B: calculated EV [2.56, 0, 1]; C: measured EV [1, 0, 6]; D: calculated EV [1, 0, 11.8]; E to F: range of the measured normals to the phase boundary (6, 0, $\bar{1}$) (8, 0, $\bar{1}$), G: calculated normal to the boundary (12.6, 0, $\bar{1}$); H: calculated position of the boundary [1, 0, 11.8]

material constants, no atomic forces, not even atomic positions entered into the calculation. This means that the result is independent of these entities, and that the same result would be obtained if the unit cells were filled with atoms in a completely different manner. The result shows, in essence—at least in this case—that the best geometrical fit is equally the most favorable from an energy point of view.

13.3. The Periodicity of the Pattern of Lattice Points

In this section we investigate the possibilities of minimum energy boundaries in the case of large differences in orientation and structure of the two crystals. Here the standard concept of the dislocation fails, but the periodicity of the pattern formed by the lattice points of both crystal lattices is considered as an essential aspect.

We have defined the 0-points as origins of the transformation $\boldsymbol{A}$ between the two lattices on the one hand, and as coincidences of elements of related equivalence classes on the other (Section 12.6). The first aspect dominates for dislocation boundaries. The 0-points within the crystal boundary are the fixed points, and origins for the relaxation between the two crystal lattices. The relaxation can be imagined to start simultaneously from all the 0-points, until it comes to a halt at the cell walls (dislocations). It is not essential to know the exact position of the 0-points as long as the 0-lattice unit cell is large compared to the crystal unit cell. It is also immaterial whether the pattern of lattice points is periodic or not.

The situation is modified, however, if the unit cell of the 0-lattice is comparable in size to that of the crystal lattice. Here it becomes important to formulate clearly the notion of the *pattern* of lattice points. We subdivide the complete pattern into *pattern elements.* A pattern element is the arrangement of crystal lattice points within one cell of the 0-lattice (bounded by the cell walls). If the complete pattern is periodic, then it consists of a finite number of different pattern elements; otherwise it consists of an infinity of them. For a given transformation $\boldsymbol{A}$ between two given lattices, the pattern element depends only on the position of the 0-point (origin) within the unit cell of lattice 1, i.e. on the internal coordinates of the 0-point. This position is determined by the related equivalence classes, the elements of which coincide at that 0-point. Thus, the number of different pattern elements is equal to the number of related equivalence classes containing coinciding elements i.e. the number of different internal coordinates of 0-points.

We shall now develop a computational method for the determination of the number of different pattern elements. For this purpose we introduce the notion of the *reduced 0-lattice,* which is understood as a single cell of lattice 1 in which all the internal coordinates of the 0-elements are plotted. (The reduced 0-lattice can contain 0-points, 0-lines or 0-planes, but here we consider only 0-points.) If the pattern is periodic, the reduced 0-lattice contains a finite number of points; otherwise this number is infinite.

Here again we come across the problem which we already met in connection with the coincidence of lattice points (Section 12.2). A pattern may be periodic, but an infinitesimal change in the relative orientation of the crystals can destroy the periodicity completely, although the physical change would be expected to be infinitesimal. Periodic and non-periodic patterns lie as close together as rational and irrational numbers.

The problem of the existence of a coincidence site lattice and of a periodic pattern are identical if at least the 0-point at the primary origin is a lattice point. However, a pattern can be periodic without containing a coincidence site lattice, if the primary origin of the transformation was not chosen as a lattice point. The analysis of the periodicity of the patterns gives a deeper insight into the meaning of the coincidence site lattice.

The relation between periodic and non-periodic patterns becomes somewhat more continuous if, instead of an infinite 0-lattice, a finite piece of it is investigated, which, for example, contains 100 0-points.

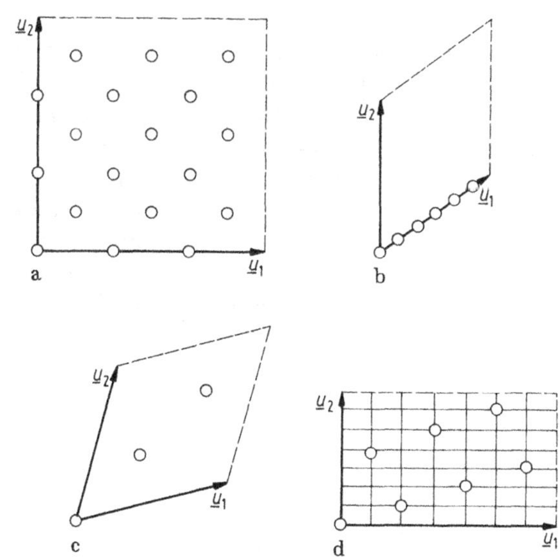

Fig. 13.3/1 a–c. Examples of reduced 0-lattices. a) Rotation {1 0 0} (Table 13.3/T 1, No. 13), b) Rotation {1 1 0} (Table 13.3/T 1, No. 31); c) Rotation {1 1 1} (Table 13.3/T 1, No. 67); d) Rotation {1 1 2} (Table 13.3/T 1, No. 72)

Its reduced 0-lattice can in this case also contain a maximum of 100 points. If, however, the period of the pattern is smaller, then the reduced 0-lattice contains less than 100 points. It is even possible that it consists of only very few points (Fig. 13.3/1); occasionally even of a single point. Starting from a highly periodic pattern consisting, for example, of two pattern elements, and changing the crystal orientation slightly so that the pattern is no longer periodic, the reduced 0-lattice will consist of 100 points, but these will now be arranged in two small groups around the original two points. However, if the observed area of the 0-lattice is extended to infinity, these small groups of points will extend, fuse and finally be distributed homogeneously over the whole unit cell of lattice 1, which houses the reduced 0-lattice. If we consider the orientation of the crystal as variable, and limit the 0-lattice to a finite range, then the fact that the reduced 0-lattice consists of small groups of points can be considered as an indication that the orientation of the crystals is close to a periodic pattern.

The search for periodic patterns (e.g. such as may be produced by the moiré models) can be performed by means of a computer in the following manner: First, 10×10 0-points are calculated in crystal coordinates, according to Eq. (12.8-5):

$$\boldsymbol{x}_{n,m}^{(0)} = n\boldsymbol{x}_1^{(0)} + m\boldsymbol{x}_2^{(0)} \qquad (n, m = 1 \ldots 10) \tag{13.3-1}$$

where $\boldsymbol{x}_i^{(0)}$ are the column vectors of the matrix $(\boldsymbol{I} - \boldsymbol{U}^{-1}\boldsymbol{S}^{-1}\boldsymbol{R}^{-1}\boldsymbol{S})^{-1}$. Of the internal coordinates of the 0-points, only the first digits after the decimal point are extracted; e.g. of the coordinates (3.725, −2.533), the internal coordinates are (0.725, 0.467). Now a 10×10 matrix $M(J, K)$ is defined in which, at the beginning, all the elements are zero. Then, the element $M(7, 4)$ is given the value 1. The same procedure is repeated for all the 100 calculated 0-points. If, in the course of the calculation, $M(7, 4)$ appears again, its value remains 1. The value of an element only indicates whether it is occupied or not, but not by how many points. The addition of the values of the matrix elements furnishes the number N of occupied positions.

$$N = \sum_{J=1}^{10} \sum_{K=1}^{10} M(J, K). \tag{13.3-2}$$

The situation corresponds to a subdivision of the reduced 0-lattice into 100 cells. When only very few positions are occupied, then the pattern is very nearly periodic. The result of such a calculation is shown in Fig. 13.3/2.

Not only is the number of pattern elements important, but also their size. The size of an 0-lattice cell is given by $1/|\boldsymbol{I} - \boldsymbol{A}^{-1}| = 1/|\boldsymbol{T}|$. Thus, the

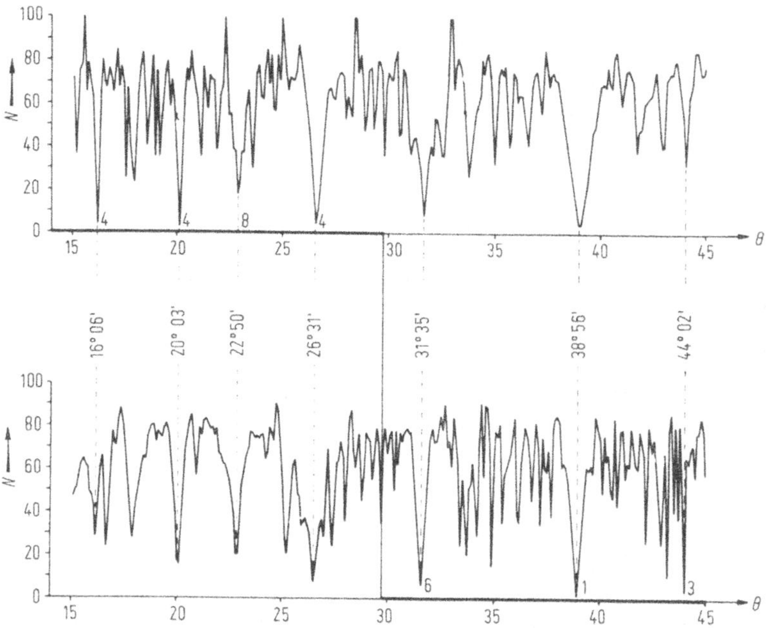

Fig. 13.3/2. Number of pattern elements $N(\theta)$ for rotation on the $\{1\,1\,0\}$-plane (compare with Fig. 12.8/4)

total period of the pattern becomes

$$N' = \frac{N}{|\boldsymbol{T}|} \,. \tag{13.3-3}$$

N' is the number of crystal units per period of the pattern. The crystal unit is the volume (the surface resp.) of the unit cell of lattice 1.

The exact values of N, N' and $|\boldsymbol{T}|$ can be determined by means of a generating function Σ in an analogous manner to that given in Section 12.2. Σ can be generalized as follows:

$$\begin{aligned} \Sigma' &= |m_1 \boldsymbol{u}_1 + m_2 \boldsymbol{u}_2|^2 \\ &= [m_1^2 |\boldsymbol{u}_1|^2 + m_2^2 |\boldsymbol{u}_2|^2 + 2 m_1 m_2 (\boldsymbol{u}_1 \boldsymbol{u}_2)] \end{aligned} \tag{13.3-4}$$

$$\Sigma = (1/\alpha)\, \Sigma' \,. \tag{13.3-5}$$

$\boldsymbol{u}_1, \boldsymbol{u}_2$ are the unit vectors of the crystal unit [column vectors of the S-matrix, Sections 2.2, 12.7, Eq. (A1-26)].

m_1, m_2 are the coordinates of the considered point of lattice 1 (according to Fig. 12.2/1).

$\alpha = 1$ or 2 depending on whether Σ' is odd or even.

The same conditions as in Section 12.2 are valid, i.e. no common factors of m_1 and m_2, and the point under investigation shall be nearest or the second nearest coincidence site to the origin. Fig. 12.2/1 shows the geometrical situation.

The practical procedure for determining periodic patterns (Table 13.3/T1) and their reduced 0-lattices consists of the following steps:

1. In a first computer programme, a table was calculated as a function of the angle of rotation θ (in steps of 0.1°). The table contained the determinant $|\boldsymbol{T}|$, the number N of elements $\neq 0$ of the matrix $M(J, K)$, $\boldsymbol{x}_1^{(0)}$ and $\boldsymbol{x}_2^{(0)}$ in crystal coordinates. Also, the values of $N' = N/|\boldsymbol{T}|$ would have been useful.

2. In a representation of the type given in Fig. 12.2/1, the value of Σ, which in most cases is identical with N', was determined. Here:

$$\theta = 2 \arctan(y/x) \quad (\text{resp. } 2 \arctan(x/y)). \qquad (13.3\text{-}6)$$

3. From that value in the table which lies closest to the θ calculated by (13.3-6), $|\boldsymbol{T}|$ is taken, and an approximate value of $N = \Sigma \cdot |\boldsymbol{T}|$ is determined. As the table was calculated only in steps of 0.1°, the calculated N might not be an exact integer, but the nearest integral value will be the correct one. The exact $|\boldsymbol{T}|$ is obtained from $N/\Sigma = N/N'$. From the coordinates of $\boldsymbol{x}_i^{(0)}$, the internal coordinates of the 0-points and hence the reduced 0-lattice can be determined.

4. In a second computer programme, on the basis of the exact θ-values, $|\boldsymbol{T}|$, the complete matrix $M(J, K)$, and the exact coordinates of the unit vectors of the 0-lattice in crystal- and in orthogonal coordinates were calculated. The matrix $M(J, K)$ gave a direct image of the reduced 0-lattice.

The difference between Σ and N' shows up if $|\boldsymbol{T}| = 0$ (and hence $N' = \infty$) while Σ remains finite. It must be stressed that Σ is a measure of the coincidence site lattice and N' a measure of the 0-lattice. The coincidence site lattice which represents the period of the pattern is exclusively determined by the pattern itself, while the 0-lattice depends on the correlation $\boldsymbol{A}$ between the two lattices. All the different 0-lattices of a given configuration contain the same coincidence site lattice (Fig. 12.8/2). In the case of $|\boldsymbol{T}| = 0$, a series of coincidence sites lie within the same 0-element. They are not separated by "dislocations". N' makes this distinction, Σ does not. Therefore, N' is a more realistic indication of the energy of a configuration.

If the pattern is represented by different transformations $\boldsymbol{A}$, then the total period (the coincidence site lattice) is independent of the actual value of $\boldsymbol{A}$. However, the transformation which correlates the nearest neighbours subdivides the total period into the smallest number of pattern elements N.

The results of the calculation for rotation on the {1 0 0}-, {1 1 0}-, {1 1 1}- and {1 1 2}-plane in the bcc structure are given in Table 13.3/T 1. This table contains the matrices $\boldsymbol{S}$, $\boldsymbol{G}$ and $\boldsymbol{U}$, the rotation angle θ, the determinant $|\boldsymbol{T}|$, the number of pattern elements N [Eq. (13.3-2)], the number of crystal units per period of the pattern N' [Eq. (13.3-3)] and the coordinates m_1, m_2 [Eq. (13.3-4)]. The unit vectors of the crystal unit in terms of the orthogonal coordinates $\boldsymbol{u}_1$ and $\boldsymbol{u}_2$ are the column vectors of the $\boldsymbol{S}$-matrix. (The length of the cube edge is taken as unit length.) The table further contains the coordinates of the points of the reduced 0-lattice. A pattern consisting of a single pattern element (coincidence site lattice) is shown in Fig. 13.3/3.

If a periodic pattern really is a minimum energy pattern, it has to be one of those with the smallest N'. These can be found in the table.

Fig. 13.3/4 shows examples of periodic patterns from the moiré models.

Table 13.3/T 1

No.	θ	$\lvert\boldsymbol{I}-\boldsymbol{A}^{-1}\rvert$	N	N'	m_1	m_2	Equivalence classes of 0-points (reduced 0-lattice)
	Rotation on {100}-plane, $\boldsymbol{S}=\boldsymbol{I}$, $\boldsymbol{G}=\boldsymbol{I}$ $\boldsymbol{U}=\boldsymbol{I}$						N defined in Eq. (13.3-2) N' defined in Eq. (13.3-3) m_1, m_2 defined in Eq. (13.3-4)
1	8° 47.8′	0.023529	2	85	13	1	(a): (0, 0) (1/2, 1/2)
2	9° 31.6′	0.027931	4	145	12	1	(b): (0,0) (0, 1/2) (1/2, 0) (1/2, 1/2)
3	10° 23.2′	0.032787	2	61	11	1	(a)
4	11° 25.2′	0.039604	4	101	10	1	(b)
5	12° 40.8′	0.048781	2	41	9	1	(a)
6	14° 15.0′	0.061539	4	65	8	1	(b)
7	16° 15.6′	0.080000	2	25	7	1	(a)
8	17° 29.2′	0.092486	16	173	13	2	(c): (0, 0) (0, 1/4) (0, 2/4) (0, 3/4); (1/4, 0) (1/4, 1/4) ...; (2/4, 0) ...
9	18° 55.4′	0.108108	4	37	6	1	(b)
10	20° 36.6′	0.128000	16	125	11	2	(c)
11	22° 37.2′	0.153846	2	13	5	1	(a)
12	25° 3.4′	0.188235	16	85	9	2	(c)
13	25° 59.4′	0.202247	18	89	13	3	(d): (0, 0) (0, 2/6) (0, 4/6); (1/6, 1/6) (1/6, 3/6) (1/6, 5/6); (2/6, 0) ...
14	28° 4.4′	0.235290	4	17	4	1	(b)
15	30° 30.6′	0.276923	18	65	11	3	(d)
16	31° 53.4′	0.301887	16	53	7	2	(c)
17	33° 24.0′	0.330275	36	109	10	3	(e): (0, 0) (0, 1/6) (0, 2/6) ... (0, 5/6); (1/6, 0) ... (1/6, 5/6); (2/6, 0) ...

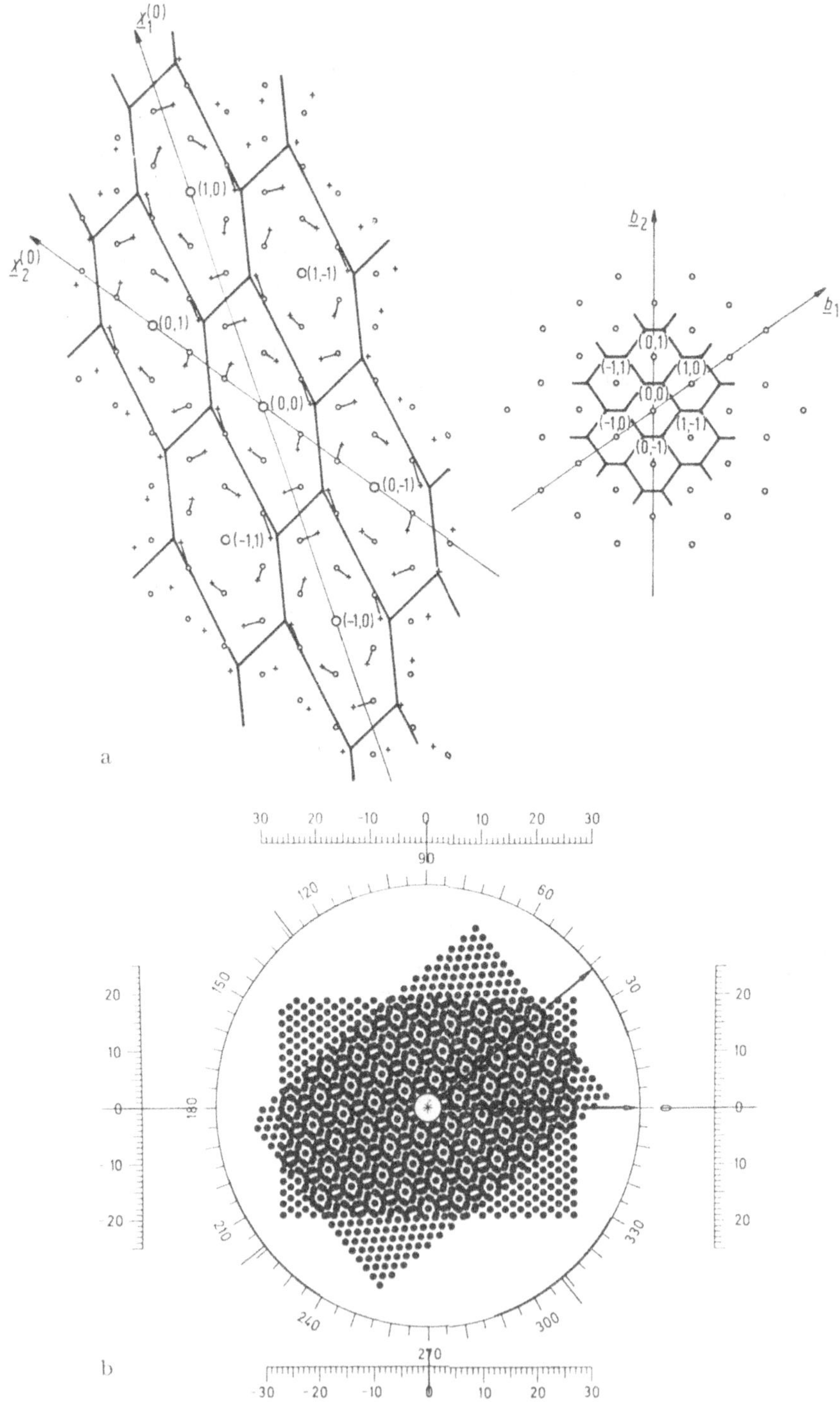

Fig. 13.3/3. a) Pattern with $N = 1$, $\{1\,1\,0\}$-rotation, $\theta = 38°\ 56.6'$ (13.3/T 1, No. 33), cell walls calculated according to Eq. (12.12-13). b) Corresponding pattern from moiré model No. 1

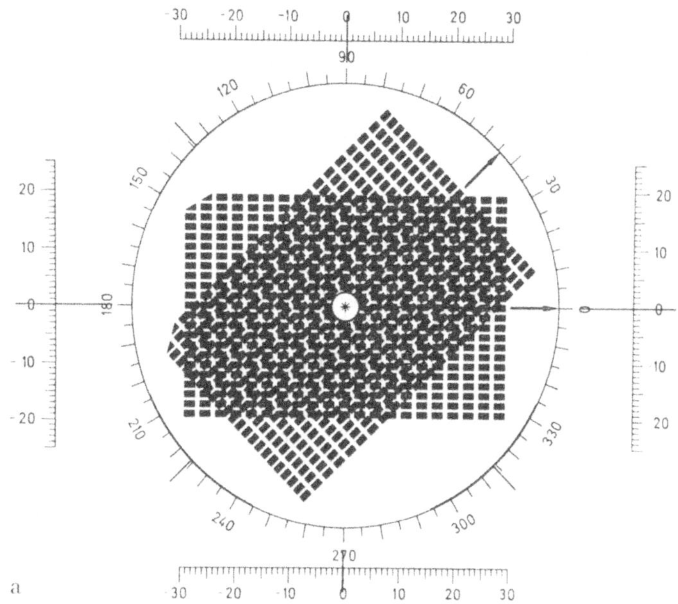

a

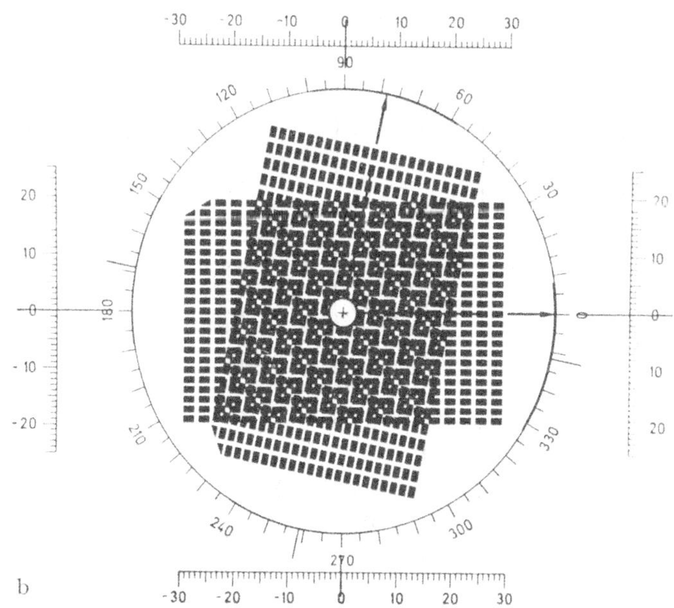

b

Fig. 13.3/4a and b. Two further examples of periodic patterns on the {1 1 2}-plane. a) Table 13.3/T1, No. 81 and b) No. 89

Table 13.3/T 1 (Continued)

No.	θ	$\lvert I-A^{-1}\rvert$	N	N'	m_1	m_2	Equivalence classes of 0-points (reduced 0-lattice)
18	34° 12.4′	0.345946	64	185	13	4	(f): (0, 0) (0, 1/8) ... (0, 7/8); (1/8, 0) ... (1/8, 7/8); (2/8, 0) ...
19	39° 52.2′	0.400000	2	5	3	1	(a)
20	39° 58.0′	0.467153	64	137	11	4	(f)
21	41° 6.8′	0.493151	36	73	8	3	(c)
22	43° 36.2′	0.551724	16	29	5	2	(c)
	Rotation on {110}-plane, $U=I$	$S=\begin{pmatrix}\sqrt{2}/2 & 0\\ 1/2 & 1\end{pmatrix}$			$G=\begin{pmatrix}3/4 & 1/2\\ 1/2 & 1\end{pmatrix}$		
23	10° 6.0′	0.031008	4	129	16	−7	(b)
24	11° 32.4′	0.040404	4	99	14	−6	(g): (0, 0) (0, 1/4) (0,2/4) (0, 3/4)
25	13° 26.6′	0.054794	4	73	12	−5	(b)
26	16° 6.0′	0.078431	4	51	10	−4	(g)
27	20° 3.0′	0.121212	4	33	8	−3	(b)
28	22° 50.4′	0.156863	8	51	14	−5	(h): (0, 0) (0, 4/8); (1/4, 3/8) (1/4, 7/8); (2/4, 2/8) (2/4, 6/8); (3/4, 1/8) (3/4, 1/8)
29	26° 31.6′	0.210526	4	19	6	−2	(g)
	$U=\begin{pmatrix}1 & 1\\ -1 & 0\end{pmatrix}$						
30	29° 42.0′	0.255474	35	137	16	−5	(0, 0) (1/35, 1/7) (2/35, 2/7) ... (6/35, 6/7); (7/35, 0) (8/35, 1/7) ...
31	31° 35.6′	0.222222	6	27	10	−3	(0, 0) (1/6, 0) (2/6, 0) ...
32	33° 43.2′	0.186916	20	107	14	−4	(0, 0); (1/20, 2/5) (3/20, 1/5) (5/20, 0); (2/20, 4/5) (4/20, 3/5) ...
33	38° 56.6′	0.111111	1	9	4	−1	(0, 0)
34	44° 0.2′	0.052632	3	57	14	−3	(i): (0, 0) (1/3, 0) (2/3, 0)
35	45° 58.6′	0.033898	2	59	10	−2	(0, 0) (1/2, 0)
36	47° 41.2′	0.019608	3	153	16	−3	(0, 0) (1/3, 0) (2/3, 0)
37	50° 28.6′	0.0		∞ (11)	6	−1	
38	53° 35.6′	−0.016260	2	123	14	−2	(0, 0) (1/2, 0)
39	55° 52.6′	−0.024390	1	41	8	−1	(0, 0)
40	58° 59.6′	−0.030303	1	33	10	−1	(0, 0)
41	61° 1.0′	−0.030928	3	97	12	−1	(0, 0) (1/3, 2/3) (2/3, 1/3)
42	62° 26.4′	−0.029851	2	67	14	−1	(0, 0) (1/2, 0)
43	83° 29.4′	−0.028249	5	177	16	−1	(0, 0) (1/5, 3/5) (2/5, 1/5) (3/5, 4/5) (4/5, 2/5)
44	70° 31.6′	0		∞ (3)	2	0	
45	79° 2.6′	0.074380	9	121	12	+1	(0, 0) (1/9, 7/9) (2/9, 5/9) ...; (5/9, 8/9) (6/9, 6/9) ...
46	80° 37.8′	0.093023	4	43	10	+1	(0, 0) (1/4, 1/2) (2/4, 0) (3/4, 1/2)
47	82° 56.8′	0.122807	7	57	8	+1	(0, 0) (1/7, 5/7) (2/7, 3/7) (3/7, 1/7) (4/7, 6/7) (5/7, 4/7) (6/7, 2/7)

Table 13.3/T1 (Continued)

No.	θ	$\lvert I-A^{-1}\rvert$	N	N'	m_1	m_2	Equivalence classes of 0-points (reduced 0-Lattice)
48	86° 37.6′	0.176471	3	17	6	+1	(0, 0) (1/3, 1/3) (2/3, 2/3)
49	89° 25.2′	0.222222	22	99	10	+2	(0, 0); (1/22, 10/11) (2/22, 9/11) ... (11/22, 0); (12/22, 10/11) ... (21/22, 1/11)
	Rotation on {111}-plane, $U=I$ $S=\begin{pmatrix}\frac{\sqrt{3}+1}{2} & \frac{\sqrt{3}-1}{2}\\ \frac{\sqrt{3}-1}{2} & \frac{\sqrt{3}+1}{2}\end{pmatrix}$ $G=\begin{pmatrix}2 & 1\\ 1 & 2\end{pmatrix}$						
50	5° 5.2′	0.007874	1	127	7	6	(0, 0)
51	6° 0.6′	0.010989	1	91	6	5	(0, 0)
52	7° 20.4′	0.016393	1	61	5	4	(0, 0)
53	9° 25.6′	0.027027	1	37	4	3	(0, 0)
54	10° 25.0′	0.032967	3	91	9	1	(j): (0, 0) (1/3, 1/3) (2/3, 2/3)
55	10° 59.4′	0.036697	4	109	7	5	(b)
56	11° 38.0′	0.041096	3	73	8	1	(j)
57	13° 10.4′	0.052632	1	19	3	2	(0, 0)
58	15° 10.6′	0.069767	3	43	6	1	(j)
59	16° 25.6′	0.081633	4	49	5	3	(b)
60	17° 53.8′	0.096774	3	31	5	1	(j)
61	19° 39.2′	0.116505	12	103	9	2	(k): (0, 0) (0, 3/6); (1/6, 1/6) (1/6, 4/6); (2/6, 2/6) (2/6, 5/6); (3/6, 0)
62	21° 47.2′	0.142857	1	7	2	1	(0, 0)
63	24° 26.0′	0.179104	12	67	7	2	(k)
64	25° 2.4′	0.187970	25	133	9	4	(l): (0, 0) (0, 1/5) (0, 2/5) (0, 3/5) (0, 4/5); (1/5, 0) (1/5, 1/5) ...
65	25° 27.6′	0.194245	27	139	10	3	(0, 0) (0, 3/9) (0, 6/9); (1/9, 1/9) (1/9, 4/9) (1/9, 7/9); (2/9, 2/9) ...
66	26° 0.4′	0.202532	16	79	7	3	(0, 0) (0, 1/4) (0, 2/4) (0, 3/4); (1/4, 0) (1/4, 1/4) ...
67	27° 47.8′	0.230769	3	13	3	1	(j)
68	23° 24.6′	0.257732	25	97	8	3	(l)
	Rotation on {112}-plane, $U=I$ $S=\begin{pmatrix}\sqrt{2} & 0\\ 0 & \sqrt{3}/2\end{pmatrix}$ $G=\begin{pmatrix}2 & 0\\ 0 & 3/4\end{pmatrix}$						
69	13° 18.4′	0.053691	8	149	1	14	(0, 0) (1/8, 1/2) (2/8, 0) (3/8, 1/2) (4/8, 0) ...
70	15° 30.0′	0.072727	4	55	1	12	(0, 0) (1/4, 1/2) (2/4, 0) (3/4, 1/2)
	$U=\begin{pmatrix}1 & 0\\ -1 & 1\end{pmatrix}$						
71	17° 24.4′	−0.091603	12	131	8	2	(0, 0) (0, 2/12) (0, 4/12) ... (0, 10/12); (1/2, 1/12) (1/2, 3/12) ...

Table 13.3/T1 (Continued)

No.	θ	$\|I-A^{-1}\|$	N	N'	m_1	m_2	Equivalence classes of O-points (reduced O-Lattice)
72	18° 33.0′	−0.090909	7	77	1	10	(0, 0) (1/7, 4/7) (2/7, 1/7) (3/7, 5/7) (4/7, 2/7) ...
73	19° 50.8′	−0.089109	9	101	7	2	(0, 0) (0, 1/3) (0, 2/3); (1/3, 0) (1/3, 1/3) (1/3, 2/3); ...
74	23° 4.4′	−0.800000	2	25	1	8	(0, 0) (0, 1/2)
75	27° 31.6′	−0.056604	3	53	5	2	(0, 0) (0, 1/3) (0, 2/3)
76	30° 27.0′	−0.034483	1	29	1	6	(0, 0)
77	34° 2.8′	0.0		∞ (35)	4	2	
78	36° 10.4′	0.024096	2	83	2	10	(0, 0) (0, 1/2)
79	36° 29.0′	0.028016	~1				
80	38° 36.8′	0.054545	3	55	7	2	(0,0) (0, 1, 1/3) (0, 2/3)
81	44° 25.0′	0.142857	1	7	1	4	(0, 0)
	$U=\begin{pmatrix}1 & 0\\-2 & 1\end{pmatrix}$						
82	52° 15.4′	−0.193548	6	31	5	4	(0, 0) (0, 1/3) (0, 2/3); (1/2, 0) (1/2, 1/3) (1/2, 2/3)
83	57° 7.2′	−0.114286	4	35	2	6	(0, 0) (0, 1/2) (1/2, 0) (1/2, 1/2)
84	62° 56.0′	0.		∞ (33)			
	$U=\begin{pmatrix}0 & 1\\-1 & 0\end{pmatrix}$						
85	66° 6.0′	−0.052829	~3	~57			
86	66° 18.2′	−0.056075	6	107	4	10	(0, 0) (1/6, 5/6) (2/6, 4/6) ... (5/6), 1/6)
87	68° 27.8′	−0.088608	7	79	5	12	(0, 0) (1/7, 6/7) (2/7, 5/7) ... (6/7, 1/7)
88	72° 35.6′	−0.142857	11	77	5	6	(0, 0) (1/11, 10/11) (2/11, 9/11) ... (10/11, 1/11)
89	78° 27.8′	−0.200000	1	5	1	2	(0, 0)
90	85° 8.4′	−0.237288	14	59	2	6	(0, 0) (1/14, 13/14) (2/14, 12/14) ... (13/14, 1/14)
91	87° 13.2′	−0.242718	25	103	7	12	(0, 0) (0, 1/5) (0, 2/5) ... (0, 4/5); (1/5, 0) (1/5, 1/5) ... (1/5, 4/5); ... (4/5, 4/5)

13.4. Displacement of the Pattern

Let us assume a minimum energy pattern as given. The question that now arises is how the crystal will behave if the relative orientation of the two lattices is slightly modified. In Section 13.3, it was shown that the periodicity of the pattern is lost. It may be energetically more favorable for the crystal to conserve the minimum energy pattern and to correct for the deviation by creating dislocations. As we know from

Section 5.4, dislocations limit areas which are displaced by translation with respect to each other. In order to be able to determine the Burgers vectors of such correcting dislocations, we have to investigate the translations of lattice 2 (and consequently those of the 0-lattice) through which the pattern as a whole is conserved.

We assume for convenience that the primary origin is a lattice coincidence site, and first restate the relations between the different translations (Sections 12.10):

$$\boldsymbol{d}^{(b)} = -\boldsymbol{A}^{-1}\,\boldsymbol{d}^{(2)} \tag{13.4-1}$$

and

$$\boldsymbol{T}\boldsymbol{d}^{(0)} = \boldsymbol{d}^{(b)}. \tag{13.4-2}$$

A pattern may be conserved in the following different ways:

1. Lattice 2 can be shifted by one of its own translation vectors. Then the arrangement of lattice points after the translation is the same as before. The b-lattice, and consequently the 0-lattice, are also shifted by one of their translation vectors. If rank$(\boldsymbol{T})$ is smaller than the rank of the problem, then, in the b-subspace, one b-lattice point is replaced by another, such that also in this case the point configuration remains unaffected by the translation. The lattice of these displacements is termed the "pattern repeat lattice" or the DR-lattice (D for displacement and R for repetition). The displacements are given by:

$$\boldsymbol{d}^{(b-R)} = \boldsymbol{b}^{(L)} \tag{13.4-3}$$

$$\boldsymbol{T}\boldsymbol{d}^{(0-R)} = \boldsymbol{b}^{(L)} \tag{13.4-4}$$

with $\boldsymbol{T} = (\boldsymbol{I} - \boldsymbol{A}^{-1})$ and:

$$\boldsymbol{d}^{(2-R)} = -\boldsymbol{A}\boldsymbol{b}^{(L)}. \tag{13.4-5}$$

2. Another possible way to conserve the pattern consists in the displacement of the primary 0-point (which is a coincidence site, and thus coincides with a lattice point of lattice 1) into another lattice point of lattice 1. The coincidence site thus appears at a new lattice point of lattice 1. The pattern as a whole is conserved but is displaced by a translation vector of lattice 1. We term the lattice of all the displacements of this type as the "elementary pattern shift lattice" or DSE-lattice, indicated by the shifts $\boldsymbol{d}^{(2-SE)}$, $\boldsymbol{d}^{(b-SE)}$ and $\boldsymbol{d}^{(0-SE)}$. The translation vectors of lattice 1 are, by definition, the $\boldsymbol{b}^{(L)}$-vectors. Hence:

$$\boldsymbol{d}^{(0-SE)} = \boldsymbol{b}^{(L)} \tag{13.4-6}$$

with

$$\boldsymbol{d}^{(b-SE)} = \boldsymbol{T}\boldsymbol{b}^{(L)} \tag{13.4-7}$$

and

$$\boldsymbol{d}^{(2-SE)} = -\boldsymbol{A}\boldsymbol{d}^{(b-SE)} = -\boldsymbol{A}\boldsymbol{T}\boldsymbol{b}^{(L)}. \tag{13.4-8}$$

3. It was shown in Section 13.3 that the reduced 0-lattice contains all the internal coordinates of the 0-points, where every point corresponds to a distinct pattern element. We may now displace the 0-point at the primary origin into every position, with internal coordinates being represented in the reduced 0-lattice. Such a displacement always leads to one of the elements of the pattern which has to be conserved, and this, in turn, leads to the complete pattern. We term the lattice of all such displacements: the "complete pattern shift lattice" or *DSC*-lattice.

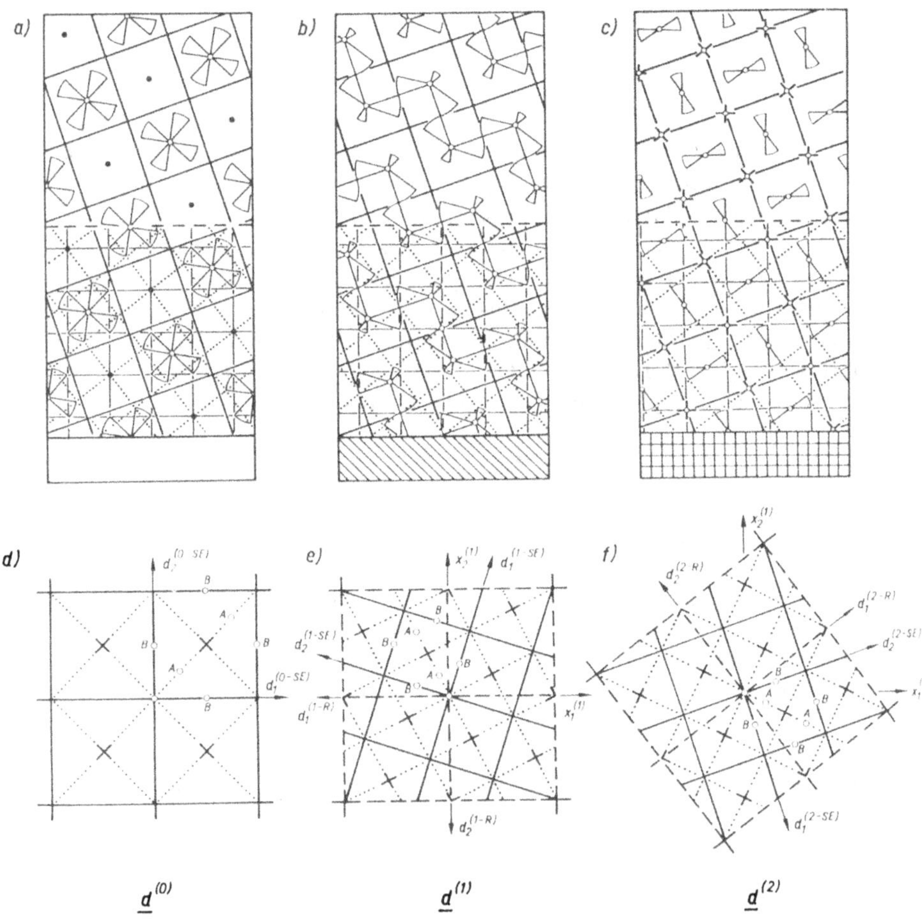

Fig. 13.4/1. a) Pattern {100}-rotation, $\theta = 36° \, 52.2'$, $N = 2$ pattern elements (13.3/T 1, No. 19); b) arising from translation of pattern (a) by $\boldsymbol{d}^{(0)} = \langle 1/2, 0\rangle$ [corresponds to points A in d), e) and f)]; c) translation $\boldsymbol{d}^{(0)} = \langle 1/2, 1/2\rangle$ (points B); d), e), f) pattern displacement lattices (enlarged 2.5 ×). Dashed lines: DR-lattice, full lines: DSE-lattice, dotted lines: DSC-lattice

Thus, the $\boldsymbol{d}^{(0-SC)}$-lattice is a configuration consisting of adjoining reduced 0-lattices (see Fig. 12.3/1). It contains, as a superlattice, the $\boldsymbol{d}^{(0-R)}$- as well as the $\boldsymbol{d}^{(0-SE)}$-lattice. The $\boldsymbol{d}^{(0-SC)}$-lattice now contains *all* the translations of the 0-lattice which conserve the pattern as a whole. It is determined by the procedure of Section 13.3. The other displacements are:

$$\boldsymbol{T}\boldsymbol{d}^{(0-SC)} = \boldsymbol{d}^{(b-SC)} \tag{13.4-9}$$

and

$$\boldsymbol{d}^{(2-SC)} = -\boldsymbol{A}\boldsymbol{d}^{(b-SC)} = -\boldsymbol{A}\boldsymbol{T}\boldsymbol{d}^{(0-SC)}. \tag{13.4-10}$$

All the displacements which deviate from the DSC-lattice modify the pattern in such a way that the reduced 0-lattice is translated within the unit cell of lattice 1. Since the reduced 0-lattice contains all the internal coordinates $0 \leqq x_i < 1$ of the 0-points, the number of points in the reduced 0-lattice remains constant upon such a translation. This means that the number of pattern elements, and consequently the total period, remains unchanged on arbitrary translations although the point arrangement of the pattern is changed, and the pattern may, for example, no longer contain coincidence sites. These features can be observed on the moiré models.

Fig. 13.4/1 presents an example of a pattern, together with the attributed DSC-lattice, as well as two patterns created by other translations. It can be reproduced by the moiré model No. 3 ($\theta = 36° 52'$) (Fig. 13.4/2).

13.5. Dislocation Networks in General Crystalline Interfaces

A dislocation network in a high angle grain boundary or a general crystalline interface will be formed if, for energetic reasons, the crystal tends to conserve a certain pattern, and if the relative orientation of the two crystals deviates only slightly from the optimum orientation. In contrast to the *primary* dislocation networks of the subgrain boundary type (intersections of the boundary with the 0-lattice cell walls), we denote the new networks as *secondary* dislocation networks. In order to calculate such a secondary dislocation network, an analogous calculation to that for the primary networks (Sections 12.9 and 12.12) is carried out. First we determine an "02-lattice", then the cell walls between the new "02-elements"; this time, however, the 02-lattice is an 0-lattice between two 0-lattices instead of between two crystal lattices.

The first 0-lattice shall have the optimum pattern given by the transformation $\boldsymbol{A}$:

$$(\boldsymbol{I} - \boldsymbol{A}^{-1})\,\boldsymbol{x}^{(0)} := \boldsymbol{T}\boldsymbol{x}^{(0)} = \boldsymbol{b}^{(L)} \tag{13.5-1}$$

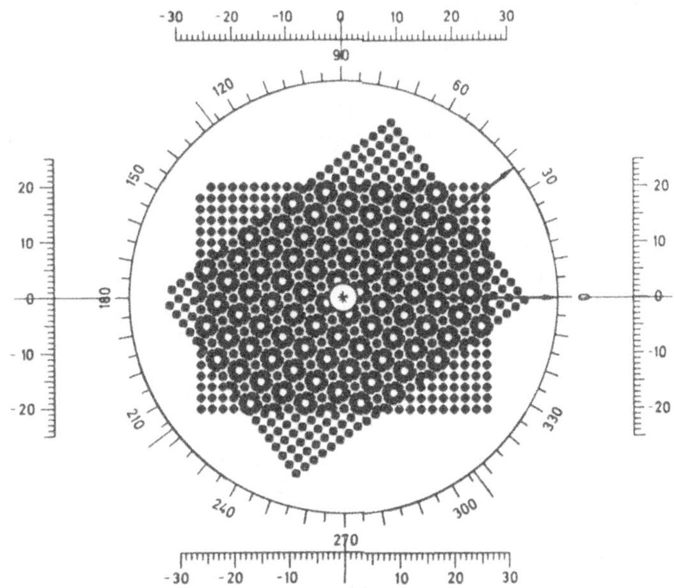

a

Fig. 13.4/2a—c. Moiré pattern corresponding to Fig. 13.4/1

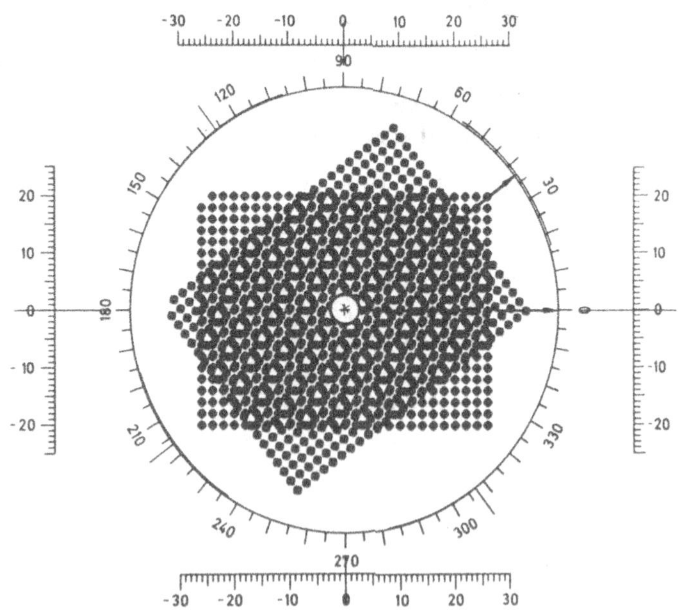

b

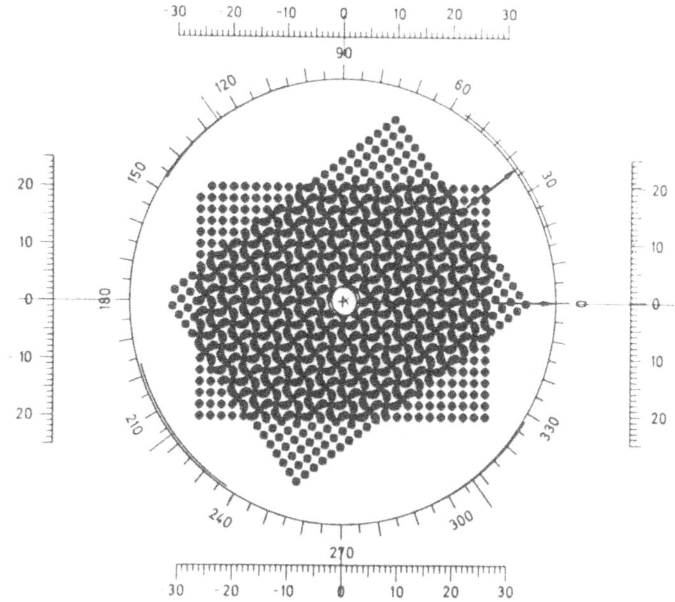

c

while the other 0-lattice shall be that which deviates slightly from the optimum pattern given by $\boldsymbol{A}'$:

$$(\boldsymbol{I}-\boldsymbol{A}'^{-1})\,\boldsymbol{x}^{(0)} := \boldsymbol{T}'\boldsymbol{x}'^{(0)} = \boldsymbol{b}^{(L)} \tag{13.5-2}$$

$\boldsymbol{x}^{(0)}$ and $\boldsymbol{x}'^{(0)}$ are images of the same $\boldsymbol{b}^{(L)}$! Eliminating $\boldsymbol{b}^{(L)}$, we obtain:

$$\boldsymbol{x}'^{(0)} = (\boldsymbol{T}'^{-1}\boldsymbol{T})\,\boldsymbol{x}^{(0)} .= \boldsymbol{B}\boldsymbol{x}^{(0)}. \tag{13.5-3}$$

Here the transformation $\boldsymbol{B}$ between two 0-lattices is analogous to the transformation $\boldsymbol{A}$ between two crystal lattices.

In analogy to the original 0-lattice, the new 0-lattice is termed the "0-lattice of second order" or "02-lattice". In order to formulate the new basic equation we have to know which vectors now correspond to the original $\boldsymbol{b}^{(L)}$-vectors. They have to be those translations of the 0-lattice which conserve the pattern as a whole i.e. the $\boldsymbol{d}^{(0-SC)}$-vectors. Thus, the basic equation becomes

$$(\boldsymbol{I}-\boldsymbol{B}^{-1})\,\boldsymbol{x}^{(02)} = \boldsymbol{d}^{(0-SC)}. \tag{13.5-4}$$

It is seen that, if $\boldsymbol{A}'=\boldsymbol{A}$ then $\boldsymbol{B}=\boldsymbol{I}$, and $(\boldsymbol{I}-\boldsymbol{B}^{-1})$ acquires rank 0, so that here no secondary dislocation network can exist.

Before we calculate the cell walls explicitely, we have to determine the Burgers vectors which are attributed to the cell walls. In a primary

O-lattice these are the differences between the $\boldsymbol{b}^{(L)}$-vectors, i.e. the translation vectors of lattice 1, which are identical with the $\boldsymbol{d}^{(b-R)}$-vectors. By analogy, the new Burgers vectors are chosen as the $\boldsymbol{d}^{(b-SC)}$-vectors, which contain the $\boldsymbol{d}^{(b-R)}$ $(=\boldsymbol{b}^{(L)})$-vectors as a subset. The conceptual problems involved with the Burgers vectors in boundaries (displacement of lattice 2 or lattice 1) will be discussed in Section 14.2.

Thus, considering the $\boldsymbol{d}^{(b-SC)}$-vectors attributed to the $\boldsymbol{x}^{(02)}$-vectors, we obtain, according to Eq. (13.4-2):

$$\boldsymbol{T}\boldsymbol{d}^{(0-SC)} = \boldsymbol{d}^{(b-SC)} \tag{13.5-5}$$

which, together with Eq. (13.5-4) gives:

$$\boldsymbol{T}(\boldsymbol{I}-\boldsymbol{B}^{-1})\,\boldsymbol{x}^{(02)} = \boldsymbol{T}\boldsymbol{d}^{(0-SC)} = \boldsymbol{d}^{(b-SC)} \tag{13.5-6}$$

with

$$\boldsymbol{T}(\boldsymbol{I}-\boldsymbol{B}^{-1}) := \boldsymbol{T}^{(02)}. \tag{13.5-7}$$

Hence, the new basic equation is:

$$\boldsymbol{T}^{(02)}\,\boldsymbol{x}^{(02)} = \boldsymbol{d}^{(b-SC)}. \tag{13.5-8}$$

Every vector $\boldsymbol{d}^{(0-SC)}$ is related on the one hand to an $\boldsymbol{x}^{(02)}$, and on the other to $\boldsymbol{d}^{(b-SC)}$, in such a way that, to every cell wall, a Burgers vector can be attributed according to the duality relations of Section 9.1.

The cell wall between the element $\boldsymbol{x}^{(02)} = [000]$ and a neighbouring $\boldsymbol{x}^{(02)}$-element is analogous to Eq. (12.12-13)

$$[\boldsymbol{d}^{(b-SC)\,T}\,\boldsymbol{G}\,\boldsymbol{T}^{(02)}]\,\boldsymbol{x} - (1/2)\,[\boldsymbol{d}^{(b-SC)\,T}\,\boldsymbol{G}\,\boldsymbol{d}^{(b-SC)}] = 0. \tag{13.5-9}$$

$\boldsymbol{G}$ is again the metric tensor of the coordinate system considered.

The Burgers vectors of the secondary dislocations in the boundary are markedly smaller than those of the primary dislocations in the single crystal. Therefore, the strain field coupled to these dislocations is also smaller.

The existence of distinct secondary dislocations compared to wavy moirés, is a proof of the existence of a minimum energy pattern. This proof was provided by Schober and Baluffi (1970). These authors soldered together gold foils, grown epitaxially on rock-salt, in a specific orientation. Fig. 13.5/1 shows an electron micrograph of foils in the {100}-plane rotated by about 36.9°. The secondary dislocations are clearly visible. Fig. 13.5/2 shows the corresponding calculated network.

Fig. 13.5/3 gives a second example of a calculated network and Fig. 13.5/4 an electron micrograph of stainless steel, but the orientation here is not known.

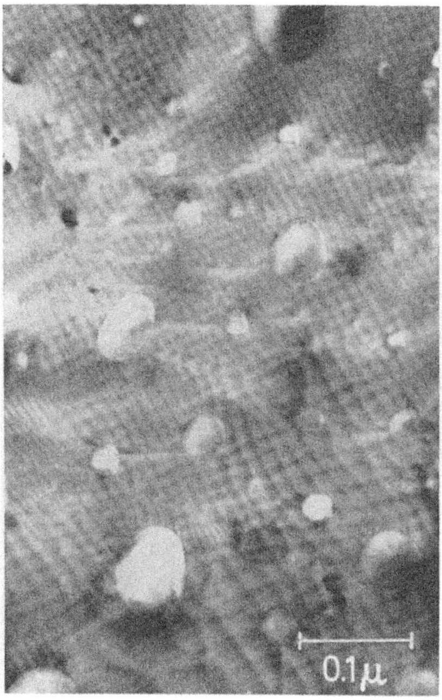

Fig. 13.5/1. Secondary dislocation network in gold corresponding to Fig. 13.5/1. (Photo: T. Schober, R. W. Balluffi)

Fig. 13.5/2a–c. Formation of a secondary dislocation network. a) Distorted pattern ({1 0 0}, $\theta = 34° \; 12.8'$). b) Secondary dislocation network conserving the minimum energy pattern Fig. 13.4/1. c) Transformation $\boldsymbol{B}$ [Eq. (13.5-3)]. d) Pattern conserving displacements of: the 0-lattice 1 and lattice 2

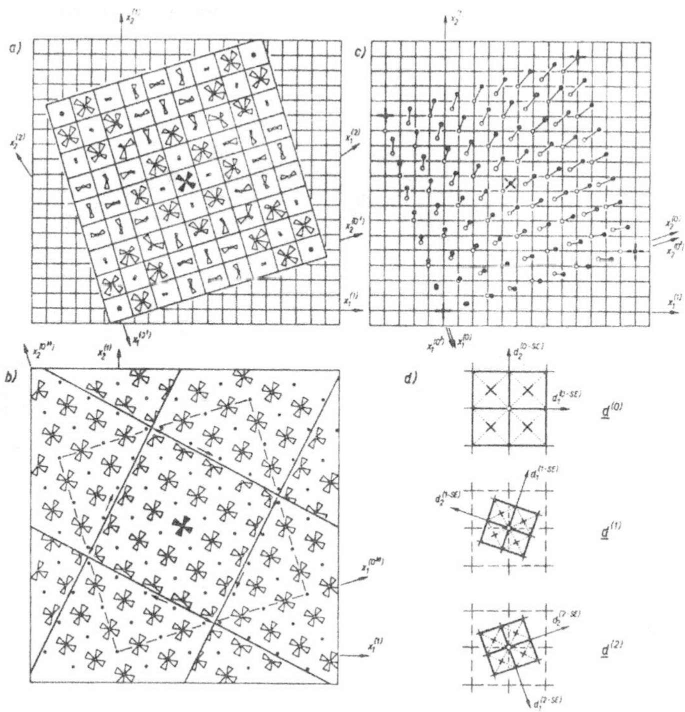

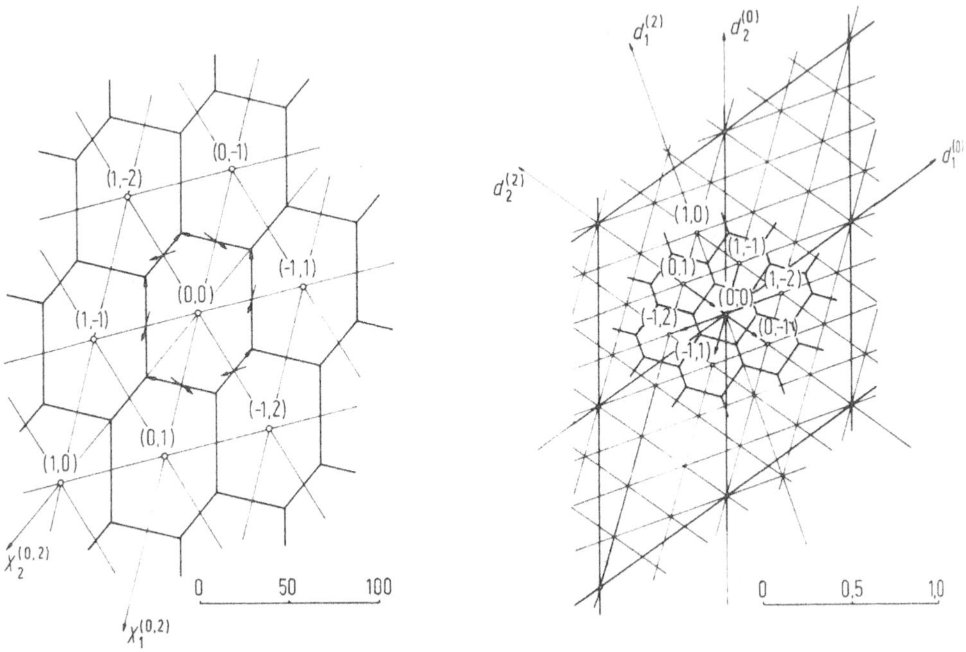

Fig. 13.5/3. Calculated secondary dislocation network, minimum energy pattern Fig. 13.3/3, $\triangle\theta = -14.6'$, to the right: $\boldsymbol{d}^{(2-SC)}$-lattice

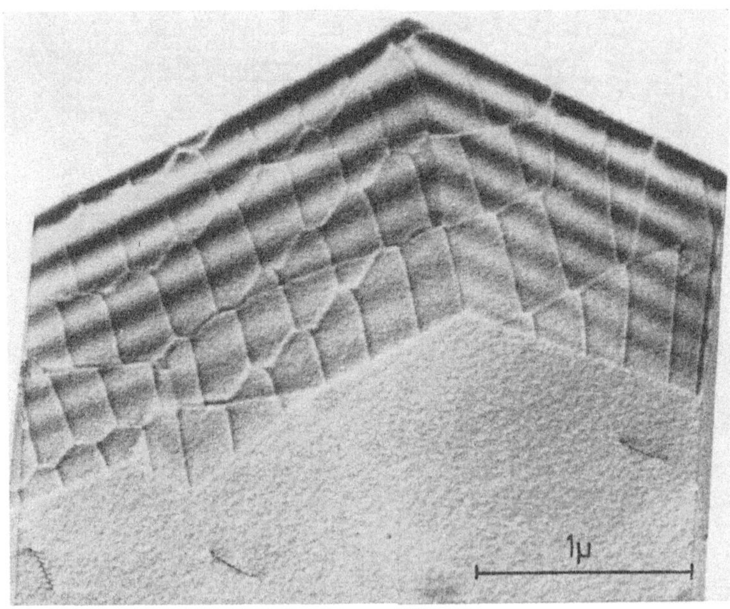

Fig. 13.5/4. Example of a secondary dislocation network. (Photo: G. Dupouy, F. Perrier)

14. Completion of the Linear 0-Lattice Theory and Extension to Non-Linear Problems

14.1. Outline

In this chapter, it will be shown how the definitions of dislocations and Burgers vectors can be extended to general crystalline interfaces. The conventional definitions of dislocations in single crystals will appear as special cases of the more general definitions. It will further be shown that practically every crystalline interface can be interpreted as a "dislocation network", and a systematics of these networks shall be presented. The three grain problem will also be outlined. With this chapter, the linear geometrical theory of crystalline interfaces is essentially closed.

Section 14.3 presents an extension of the theory to non-linear problems which arise when a dislocation from the interior of the crystal ends at a boundary or penetrates it.

14.2. Dislocations and Burgers Vectors in Crystalline Interfaces

In Section 5.2 the dislocation line was defined by a Volterra cut. Another possibile definition followed from the movement rule (Section 5.4) where the dislocation line was understood as being a boundary line between a displaced and an undisplaced region.

The definition and sign of the Burgers vector followed accordingly from the Burgers loop or the displacement during the movement. In connection with the Burgers loop, it was emphasized that the Burgers vector has to be conceived as a coordinate difference in the (eventually elastically distorted) coordinate system of the crystal, in order that its invariance along the line and its continuity on dislocation branching are preserved. This fact was already taken account of by Frank by introduction of the reference lattice. A clear distinction is given in this way between plastic and elastic deformation.

The dislocations within a crystal boundary are defined as the intersections of the boundary surface with the 0-lattice cell walls (Section 13.1). The Burgers vectors are attributed to the dislocations by means of the duality relations between the dislocation network and its b-net (Section 9.1 and 11.2).

We shall now investigate how far the original definitions (Burgers loop, movement rule) remain valid for dislocations in crystal boundaries, and what aspects of the notion of a dislocation still have a meaning in relation to general crystalline interfaces. For this purpose we outline the analytical representation of a Burgers loop.

We start, for example, in crystal 1 in the cell of the 0-point $\boldsymbol{x}^{(0)}$ at $\boldsymbol{x}^{(1LS)}$ (S = start) (Fig. 14.2/1 a) and proceed inside that cell by translation vector steps to the crystal boundary $\boldsymbol{x}^{(1L)}$. At the boundary we jump to the related point in lattice 2, $\boldsymbol{x}^{(2L)}$, where:

$$(\boldsymbol{x}^{(2L)} - \boldsymbol{x}^{(0)}) = \boldsymbol{A}\,(\boldsymbol{x}^{(1L)} - \boldsymbol{x}^{(0)})\,. \qquad (14.2\text{-}1)$$

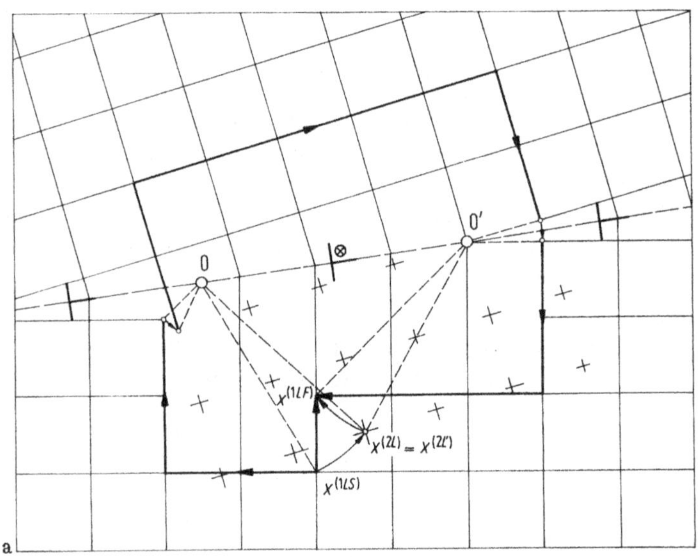

a

Fig. 14.2/1 a and b. Burgers loop across a high-angle grain boundary. a) Start and end of the loop in crystal 1, b) in crystal 2

From the point $\boldsymbol{x}^{(2L)}$ we continue in translation vector steps, now in lattice 2, into the cell of the neighbouring 0-points $\boldsymbol{x}'^{(0)}$, and from there we return over the boundary to the related point in lattice 1, which is now defined with respect to $\boldsymbol{x}'^{(0)}$:

$$(\boldsymbol{x}'^{(1L)} - \boldsymbol{x}'^{(0)}) = \boldsymbol{A}^{-1}(\boldsymbol{x}'^{(2L)} - \boldsymbol{x}'^{(0)})\,. \qquad (14.2\text{-}2)$$

We then continue inside lattice 1 to the endpoint $\boldsymbol{x}^{(1LF)}$ (F = finish). We may use "pro memoria" the reference lattice inside which the loop has to be closed (Section 5.2).

If the translation vectors in lattice 2 are considered as being equivalent to those in lattice 1 through

$$\boldsymbol{t}^{(2)} = \boldsymbol{A}\boldsymbol{t}^{(1)} \tag{14.2-3}$$

it can be shown that every single loop, however complicated it may be, can be reduced to the following steps (Fig. 14.2/1a):

$$\boldsymbol{x}^{(1LS)} \xrightarrow{(0)} \boldsymbol{x}^{(2L)} \longrightarrow \boldsymbol{x}'^{(2L)} \xrightarrow{(0')} \boldsymbol{x}^{(1LF)}. \tag{14.2-4}$$

In general,

$$\boldsymbol{x}'^{(2L)} - \boldsymbol{x}^{(2L)} = 0 \quad \text{i.e.} \quad \boldsymbol{x}^{(2L)} = \boldsymbol{x}'^{(2L)}. \tag{14.2-5}$$

The exceptions are those cases discussed in Section 12.11, where

$$\boldsymbol{x}'^{(2L)} - \boldsymbol{x}^{(2L)} = \boldsymbol{A}(\boldsymbol{b}'^{(L\perp)} - \boldsymbol{b}^{(L\perp)}). \tag{14.2-6}$$

Now, the sequence (14.2-4) will be formulated in detail:

$$\boldsymbol{x}^{(2L)} - \boldsymbol{x}^{(0)} = \boldsymbol{A}(\boldsymbol{x}^{(1LS)} - \boldsymbol{x}^{(0)}) \tag{14.2-7}$$

$$\boldsymbol{x}^{(2L)} - \boldsymbol{x}'^{(0)} = \boldsymbol{A}(\boldsymbol{x}^{(1LF)} - \boldsymbol{x}'^{(0)}). \tag{14.2-8}$$

When $\boldsymbol{x}^{(2L)}$ is eliminated, we obtain:

$$(\boldsymbol{I} - \boldsymbol{A}^{-1})(\boldsymbol{x}'^{(0)} - \boldsymbol{x}^{(0)}) = \boldsymbol{x}^{(1LF)} - \boldsymbol{x}^{(1LS)}. \tag{14.2-9}$$

Taken together with the basic Eq. (12.3-6), this gives:

$$\boldsymbol{x}^{(1LF)} - \boldsymbol{x}^{(1LS)} = \boldsymbol{b}'^{(L)} - \boldsymbol{b}^{(L)}. \tag{14.2-10}$$

This means that the closure failure of the loop is equal to the difference of the b-lattice vectors, the images of which are $\boldsymbol{x}^{(0)}$ and $\boldsymbol{x}'^{(0)}$. It should be noted that, in the disturbed crystal, the Burgers vector has to point from the beginning to the end of the loop.

If the loop starts in crystal 2, then the steps are (Fig. 14.2/1b):

$$\boldsymbol{x}^{(2LS)} \xrightarrow{(0')} \boldsymbol{x}'^{(1L)} \longrightarrow \boldsymbol{x}^{(1L)} \xrightarrow{(0)} \boldsymbol{x}^{(2LF)}. \tag{14.2-11}$$

(Attention has to be paid to the sequence of 0′ and 0, in order that the sense of the loop is preserved!)

The result here is:

$$\boldsymbol{A}^{-1}(\boldsymbol{x}^{(2LF)} - \boldsymbol{x}^{(2LS)}) = \boldsymbol{b}'^{(L)} - \boldsymbol{b}^{(L)}. \tag{14.2-12}$$

The steps in lattice 2 are equivalent to those in lattice 1, insofar as both lattices are coupled through the transformation $\boldsymbol{A}$.

In the case of rank$(\boldsymbol{T}) < 3$, the following holds:

$$\boldsymbol{x}'^{(1L)} - \boldsymbol{x}^{(1L)} = \boldsymbol{b}'^{(L\perp)} - \boldsymbol{b}^{(L\perp)} \tag{14.2-13}$$

which has to be taken into account in the loop steps (14.2-11).

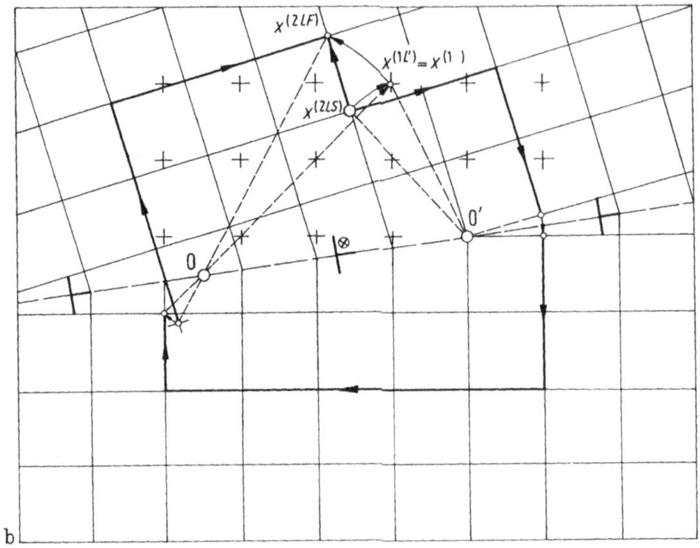

Fig. 14.2/1 b

It has been shown above that the definition of the Burgers vectors obtained from the duality relation between the 0- and the b-lattices is equivalent to the original definition by the Burgers loop, if translation vectors related by $\boldsymbol{A}$ in both lattices are taken as being "identical".

The kinematic definition of the Burgers vector (movement rule, Section 5.4) can also be extended to general crystalline interfaces. For the case of a single dislocation in a perfect crystal it was assumed that the dislocation could be displaced by arbitrary amounts in the direction $\boldsymbol{v}$. In the present context, we make restrictions to this assumption. The dislocation shall be displaced by one 0-lattice cell period, such that an 0-element after the displacement takes the position of a neighbouring element. The total pattern before and after the displacement is identical with respect to structure *and* position.

This is achieved either by keeping lattice 1 fixed and displacing lattice 2 by one of its minimum translation vectors [$\boldsymbol{d}^{(2-R)}$ Eq. (13.2-3) to (13.2-5)], or vice versa by keeping lattice 2 fixed and shifting lattice 1 by its corresponding translation vector $\boldsymbol{d}^{(1-R)} = \boldsymbol{d}^{(b-R)}$. If $\boldsymbol{m}$ [of the movement rule Eq. (5.4-1)] points into lattice 2, then $\boldsymbol{d}^{(2-R)}$ is the translation vector. If the direction and size of $\boldsymbol{v}$ are left unchanged then $\boldsymbol{m}$ can only change its sign, and thus point into crystal 1 (which now becomes the shifted part), if $\boldsymbol{b}$ and $\boldsymbol{l}$ simultaneously change their sign. In addition, $\boldsymbol{b}$ (i.e. $\boldsymbol{d}^{(R)}$) changes its orientation, and eventually its size, on the transition from lattice 2 into lattice 1, i.e.

$$\boldsymbol{d}^{(1-R)} = -\boldsymbol{A}^{-1}\,\boldsymbol{d}^{(2-R)} \tag{14.2-14}$$

where

$$\boldsymbol{d}^{(1-R)} = \boldsymbol{b}'^{(L)} - \boldsymbol{b}^{(L)}. \tag{14.2-15}$$

Thus, also here the Burgers vector is the *coordinate difference* of minimum translation vectors in both lattices which are coupled through $\boldsymbol{A}$ (minimum, because it was assumed that the 0-element is shifted into the neighbouring position).

After having seen that the notion of the dislocation can be extended to general crystalline interfaces, these interfaces can also be interpreted as "dislocation networks". (An exception is the perfect twin boundary which is formed by an 0-plane.) We distinguish on the one hand *physical-* and *geometrical* (or mathematical) dislocations and on the other hand *primary* and *secondary* dislocations.

We understand here as a dislocation line the intersection of the crystal boundary with a lattice cell wall. It is a *physical* dislocation if the 0-lattice cell is large compared to the unit cell of the crystal lattice. In this case, the relaxation which starts from the 0-elements acts such that conventional dislocations, with cores and strain fields, are formed at the intersection of the boundary with the cell walls.

However, if the 0-lattice cell is of similar size to the crystal unit cell, then no physical dislocation can be formed any more, but, by definition, the intersection of the boundary with a cell wall is also here a "dislocation" with the invariance and continuity properties of the Burgers vector (the latter follows from the existence of a connected b-net.) The "dislocation" is now a geometrical delimitation of the different 0-element regions, and as such it is called a *geometrical* dislocation.

The distinction between primary and secondary dislocations can be seen best from the behaviour of the pattern on translation of one crystal lattice by a moving dislocation. Is the pattern preserved with respect to structure *and* position, the translation is caused by a (perfect) *primary* dislocation, be it physical or geometrical. If, however, the pattern is preserved only with respect to its structure but is allowed to change its position, then the translation is caused by a *secondary* dislocation. As was shown in Section 13.4, the Burgers vectors of the primary dislocations form a subset of the Burgers vectors of the secondary dislocations (the $\boldsymbol{d}^{(b-R)}$ correspond to the primary Burgers vectors and these are included in the $\boldsymbol{d}^{(b-SC)}$ of the secondary Burgers vectors).

Secondary dislocations are formed if a minimum energy pattern has to be preserved. Each one of these patterns has its own DSC-lattice (Section 13.4) and thus its own set of secondary Burgers vectors. The physics of the crystal will finally determine what secondary dislocations are formed.

Fig. 14.2/2 presents schematically the ranges of the different types of dislocation networks. The path of the energy curve corresponds formally to that of the subgrain boundary (Section 11.7). The primary dislocations extend over the whole range. With the exception of the

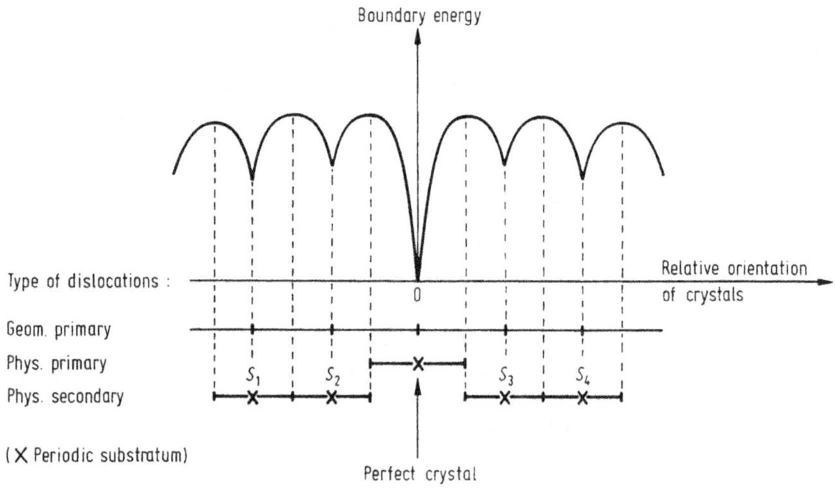

Fig. 14.2/2. Schematic representation of the ranges of different types of dislocations

surroundings of the perfect crystal, they have to be considered as geometrical dislocations. If the transformation $\boldsymbol{A}$ is transferred to a new relation (Section 12.8), then the structure of the primary dislocation network is also changed. The secondary dislocation networks are attributed to their corresponding minimum energy patterns.

The situation of the primary and the secondary dislocations can also be characterized in the following way: We assume an arrangement of two crystals with a boundary. From the standpoint of primary dislocations this arrangement corresponds to one crystal containing a dislocation network; from the standpoint of secondary dislocations, however, it can be regarded as a sandwich of three "crystals": crystal 1, the periodic pattern of the boundary, and crystal 2. The periodic pattern in the boundary acts as a periodic substrate for its own set of possible dislocations. It will be shown in the next section that the dislocation balance is also fulfilled, in most cases, if a dislocation from a crystal enters or penetrates a boundary.

The 0-lattice procedure can be extended to three grains as follows: We assume that three crystals grains are given in known orientations. Lattices 1 and 2 are then related by

$$\boldsymbol{x}^{(2)} = \boldsymbol{A}^{(21)}\,\boldsymbol{x}^{(1)} \tag{14.2-16}$$

(where $\boldsymbol{A}^{(21)}$ relates the nearest neighbours), and lattices 1 and 3 by

$$\boldsymbol{x}^{(3)} = \boldsymbol{A}^{(31)}\,\boldsymbol{x}^{(1)}. \tag{14.2-17}$$

On eliminating $\boldsymbol{x}^{(1)}$ we obtain:

$$\boldsymbol{x}^{(3)} = [\boldsymbol{A}^{(31)}(\boldsymbol{A}^{(21)})^{-1}]\,\boldsymbol{x}^{(2)} := \boldsymbol{A}^{(32)}\,\boldsymbol{x}^{(2)}. \tag{14.2-18}$$

$\boldsymbol{A}^{(32)}$ has to be checked and eventually corrected by a unimodular transformation by the procedure given in Section 12.8 in order to again correlate the nearest neighbours. On the basis of $\boldsymbol{A}^{(32)}$ the 0-lattice and the boundary between crystal 2 and 3 can be determined.

14.3. Non-Linear Problems

Up till now we have regarded lattice 1 and lattice 2 as translation lattices and developed the 0-lattice theory on that basis. Now we shall extend the procedure to non-linear problems, e.g. to distorted lattices or lattices containing dislocations. For this purpose we represent the two "lattices" as functions of a vector parameter $\boldsymbol{z}$ $\left(\boldsymbol{x}^{(1)}(\boldsymbol{z})\right.$ and $\left.\boldsymbol{x}^{(2)}(\boldsymbol{z})\right)$. We assume the functions to be essentially continuous (curvilinear coordinates) and the "lattice points" shall be determined by integral values of $\boldsymbol{z}$. Here we denote the "period" of both lattices by the integral vectors $\boldsymbol{b}^{(L)}$. The lattices, partitioned into equivalence classes, are now represented by the "lattice" of the $\boldsymbol{z}$-vectors. The condition for the coincidence of elements of related equivalence classes becomes:

$$\underline{\boldsymbol{x}^{(2)}(\boldsymbol{z}) = \boldsymbol{x}^{(1)}(\boldsymbol{z} + \boldsymbol{b}^{(L)})}. \tag{14.3-1}$$

From this equation the intermediate solution:

$$\boldsymbol{z} = \boldsymbol{z}(\boldsymbol{b}^{(L)}) \tag{14.3-2}$$

is determined. $\boldsymbol{x}^{(0)}$ is then given either by

$$\underline{\boldsymbol{x}^{(0)}(\boldsymbol{b}^{(L)}) = \boldsymbol{x}^{(2)}\left(\boldsymbol{z}(\boldsymbol{b}^{(L)})\right)} \tag{14.3-3}$$

or by

$$\underline{\boldsymbol{x}^{(0)}(\boldsymbol{b}^{(L)}) = \boldsymbol{x}^{(1)}\left(\boldsymbol{z}(\boldsymbol{b}^{(L)}) + \boldsymbol{b}^{(L)}\right)}. \tag{14.3-4}$$

We can show that this procedure is a generalization of the basic Eq. (12.3-6).

We set:

$$\boldsymbol{x}^{(1)}(\boldsymbol{z}) = \boldsymbol{x}^{(1)} \tag{14.3-5}$$

$$\boldsymbol{x}^{(1)}(\boldsymbol{z} + \boldsymbol{b}^{(L)}) = \boldsymbol{x}^{(1)} + \boldsymbol{b}^{(L)} \tag{14.3-6}$$

and

$$\boldsymbol{x}^{(2)}(\boldsymbol{z}) = \boldsymbol{A}\boldsymbol{x}^{(1)} \tag{14.3-7}$$

which again leads to the basic equation.

An example will now be discussed of a dislocation which ends at a grain boundary. The boundary is assumed to be the one formed by a rotation of 36° 52.2′ in the cubic {100}-plane. The pattern is given in Fig. 13.4/1a with the number of pattern elements $N=2$ (coincidence site and "cross"). Its period is $N'=5$ crystal units. Again we let two lattices interpenetrate, one containing an edge dislocation, the other being perfect.

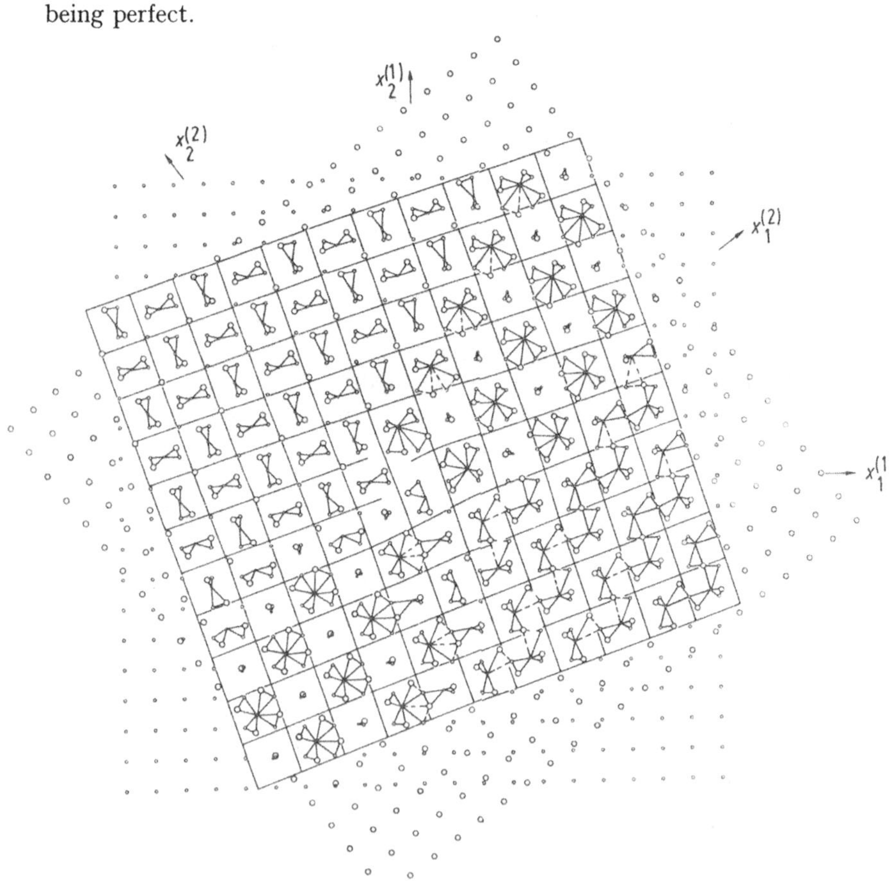

Fig. 14.3/1. Calculated pattern for the case of a dislocation ending at a grain boundary (original pattern Fig. 13.4/1a)

The strain field of an edge dislocation is, according to Section 6.6, given by:

$$u_1(x_1, x_2) = \frac{b}{2\pi}\left[\arctan\left(\frac{x_2}{x_1}\right) + \frac{1}{2(1-\nu)}\,\frac{x_1 x_2}{x_1^2+x_2^2}\right]$$
$$u_2(x_1, x_2) = \frac{-b}{8\pi(1-\nu)}\left[(1-2\nu)\cdot\ln(x_1^2+x_2^2) + \frac{x_1^2-x_2^2}{x_1^2+x_2^2}\right]. \qquad (14.3\text{-}8)$$

The crystal containing the edge dislocation (here crystal 2) is represented by $\boldsymbol{x}+\boldsymbol{u}(\boldsymbol{x})$, where $\boldsymbol{x}$ corresponds to the parameter $\boldsymbol{z}$. Lattice 2 shall be rotated by $\boldsymbol{R}$ while the dislocation-free lattice 1 determines the coordinate system. In this manner the generalized basic equation becomes:

$$\boldsymbol{R}\left(\boldsymbol{x}+\boldsymbol{u}(\boldsymbol{x})\right) - (\boldsymbol{x}+\boldsymbol{b}^{(L)}) = 0. \qquad (14.3\text{-}9)$$

Written in detail the equations are:

$$r_{11}[x_1+u_1(x_1\,x_2)] + r_{12}[x_2+u_2(x_1\,x_2)] - (x_1+b_1^{(L)}) = 0,$$
$$r_{21}[x_1+u_1(x_1\,x_2)] + r_{22}[x_2+u_2(x_1\,x_2)] - (x_2+b_2^{(L)}) = 0. \qquad (14.3\text{-}10)$$

$\boldsymbol{x}(\boldsymbol{b}^{(L)})$ and then $\boldsymbol{x}^{(0)} = \boldsymbol{x}(\boldsymbol{b}^{(L)}) + \boldsymbol{b}^{(L)}$ were calculated and the result is given in Fig. 14.3/1. Here the 0-lattice itself contains a "dislocation". Thus, we have calculated the configuration which appears when a dislocation branches out of a network and for which in Section 11.9 the duality relations failed. It is to be stressed that, if the intersection lines of the crystal boundary with cell walls are strictly understood as "primary" dislocations, the Burgers vector balance (node condition, Section 7.4) is closed over the whole network.

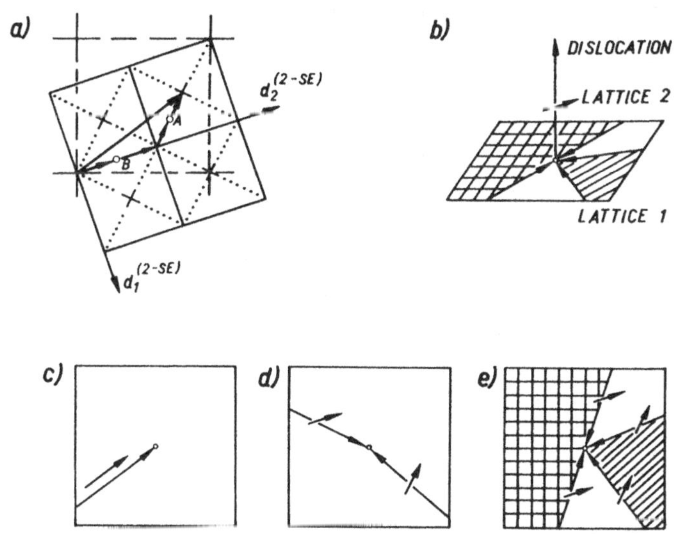

Fig. 14.3/2. Interpretation of Fig. 14.3/1 in terms of secondary dislocations (Hatching: see Fig. 13.4/1)

The b-lattice can be formally looked on as a cubic lattice, but the relation between the b- and the 0-lattices is no longer given by a linear transformation.

The situation can also be interpreted from the standpoint of the secondary dislocations (Fig. 14.3/2): the incoming dislocation is first split into two secondaries and these again are split into secondary "partials". This second splitting follows from purely geometrical considerations. The energy situation will decide for each individual case whether a dislocation will split or not.

Fig. 14.3/3 shows the case where an edge dislocation penetrates a boundary. Here both lattices contain an edge dislocation and the basic

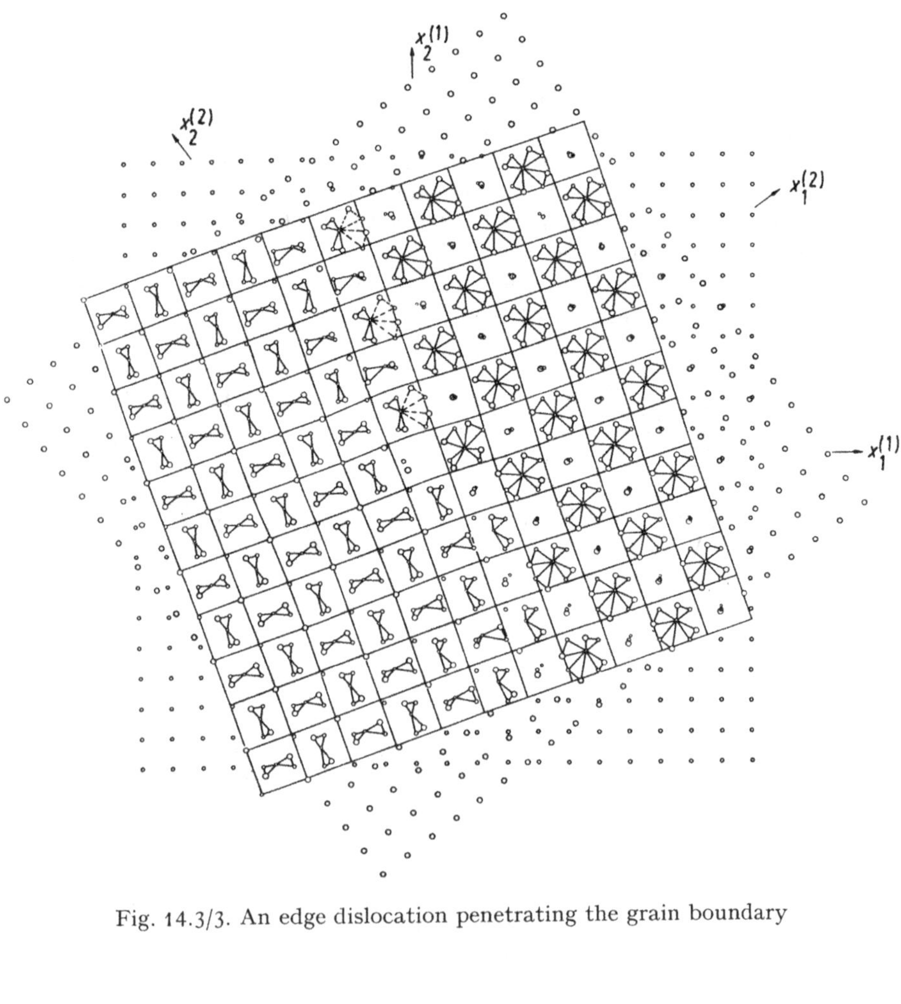

Fig. 14.3/3. An edge dislocation penetrating the grain boundary

equation becomes:

$$\boldsymbol{R}(\boldsymbol{x}+\boldsymbol{u}(\boldsymbol{x}))-[\boldsymbol{x}+\boldsymbol{b}^{(L)}+\boldsymbol{u}(\boldsymbol{x}+\boldsymbol{b}^{(L)})]=0. \tag{14.3-11}$$

From the standpoint of the primary dislocations the boundary is a somewhat deformed square network. No dislocation branches off. The dislocations entering and leaving the boundary in the two different crystals are equivalent, and their Burgers vectors (coordinate differences!) cancel at the boundary although the corresponding translation vectors in both lattices are rotated with respect to each other by about 37°. From the standpoint of the secondary dislocations, however, a dislocation branches off in the boundary and appears again as split into partials (Fig. 14.3/4); hence, from this view point the two crystal dislocations are no longer equal.

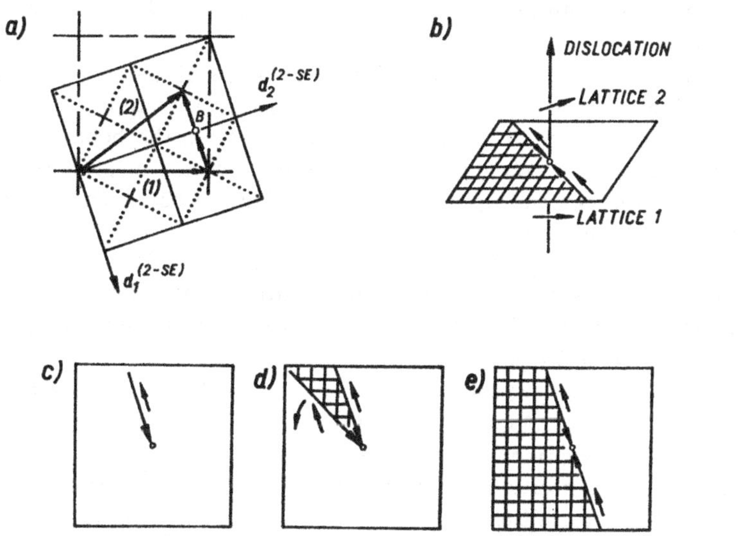

Fig. 14.3/4. Interpretation of Fig. 14.3/3 in terms of secondary dislocations

In the case given in Fig. 14.3/3 an edge dislocation penetrates a perfect "minimum energy boundary" (optimum boundary). The difference between the outgoing and the incoming translation vectors is exactly equal to the Burgers vector of a secondary dislocation (Fig. 14.3/4a). If, however, the rotation between the crystal lattices does not exactly correspond to an optimum boundary, then a network of secondary dislocations is formed in the boundary (Section 13.5). The secondary dislocation which branched off will be woven into that network. Now the difference vector between the corresponding translation vectors of lattice 2 and lattice 1 will not be an exact Burgers vector of a secondary dislocation. The nearest secondary — and eventually primary Burgers vector is still a certain difference away. This difference does not fit into the Burgers vector balance, but it exerts a strain similar to that of a dislocation with a *rest*-Burgers vector. Marcinkowski (1970) terms such a rest-dislocation a *virtual grain boundary dislocation* (abbreviated to VGBD). A virtual dislocation also appears in a *low angle* boundary when it is penetrated by a dislocation. In such a situation, the minimum energy pattern is the pattern of the perfect crystal, and the possible Burgers vectors are its translation vectors. Similar problems have been discussed by Sleeswyk (1967). He introduced the notion of an unstable dislocation, which is somehow related to the virtual dislocation. To summarize, a virtual dislocation can appear if a boundary is penetrated by a dislocation. In the case where the dislocation ends at the boundary, the balance is always closed although the *b*-net is no longer topologically planar.

Until now we have assumed that an edge dislocation ends at the boundary or penetrates it. The same can naturally happen with a screw dislocation. If the dislocation does not penetrate the boundary it continues inside the boundary as an edge dislocation and produces a step. Such a step can end up as a spiral whereby the boundary is displaced. Spirals in boundaries were observed by Gleiter (1969a). Gleiter (1969b) developed a theory of grain boundary motion based on a model where the two grains limited by steps are separated by some interlayer. The grain boundary motion takes place by diffusion of atoms in different stages through the interlayer.

If on the basis of the 0-lattice theory, a model of the grain boundary motion were developed, the atoms sitting in lattice 1 would have to shift into a position in lattice 2. It is most probable that an atom would pass from its position in lattice 1 into the nearest position in lattice 2 (i.e. into that position which is related to position 1 by $\boldsymbol{A}$, with respect to the nearest 0-point taken as origin). These two positions are usually less than one atomic diameter apart, so that, on the average, the actual displacement of the individual atoms would be of the same order of magnitude. No interlayer would be needed in this description. The atoms would not have to pass through the boundary, the boundary would pass through the atoms.

Chapters 13–14. Summary and Discussion

Some applications of the 0-lattice theory are given in chapters 13 and 14. Two different cases have to be distinguished:

1. the unit cell of the 0-lattice is large compared to that of the crystal lattice, and

2. the unit cell of the 0-lattice and of the crystal lattice are of similar dimensions.

The first case leads to boundaries which consist of dislocation networks of the subgrain boundary type. An important problem is the determination of "optimum" boundaries where the orientation of the crystals, as well as the orientation of the boundary, have to be determined in such a way that the value of the boundary energy attains a minimum. No optimum sub*grain* boundaries exist because the "optimum" state is represented by the perfect crystal. It is possible, however, to find optima for phase boundaries. An example is discussed in Section 13.2. If the problem is to *locate* an energy minimum without expecting quantitative indications on the energy values, it is sufficient to replace the energy by a function which can be anticipated to vary monotonically with the energy.

The dislocation lines can be defined as the lines of intersection of the crystal boundary with the 0-lattice cell walls. Starting from the different 0-points as origins, the atoms in the crystals will align themselves by relaxation until, at the cell wall, the coordination changes; this change is the dislocation. For "physical" dislocation networks it is not essential to locate exactly the 0-points, i.e. to know, for example, their internal coordinates.

However, for high-angle boundaries or boundaries between widely different phases the situation is modified (Section 13.3). In this case optimum boundaries are possible too, but on a completely different basis. The decisive factor is the periodicity of the pattern of lattice points. Here, the total pattern is subdivided by the 0-lattice cells into pattern elements, and for a given transformation $\boldsymbol{A}$ the pattern element is determined by the internal coordinates (the equivalence class) of its 0-point. The essential data are the number of equivalence classes with coinciding elements, the internal coordinates of the 0-points, and the size of the 0-lattice cell. The smallest values of Eq. (13.3-3) should correspond to an optimum boundary; the period of the pattern is independent of translation and so is this equation. Yet, it might be expected that, out of all the patterns which differ only by translation, the most stable pattern would be that in which the 0-points are occupied by atoms.

Once again the difference between the 0-lattice and the coincidence site lattice should be emphasized: the coincidence site lattice is determined exclusively by the point pattern and is independent of the relation $\boldsymbol{A}$ between the two lattices. It indicates the period of the pattern and is discontinuous with respect to any change in $\boldsymbol{A}$. Hence, "coincidence site lattice" is synonymous with "periodic pattern". The 0-lattice, however, depends strictly on $\boldsymbol{A}$. It varies continuously with $\boldsymbol{A}$ and, if a coincidence site lattice is present, it is contained in the 0-lattice. There are cases where the 0-lattice and the coincidence site lattice are identical. However, the information content of the coincidence site lattice is only a small fraction of that of the 0-lattice.

Slight deviations from optimum patterns can lead to secondary dislocation networks. When secondary dislocations move through the boundary the structure of the optimum pattern is conserved, but not necessarily its position (Section 13.5).

The whole linear geometrical theory of crystalline interfaces is, in a sense, condensed in Section 14.2. Practically every boundary can be understood as a "dislocation network" where, however, only a geometrical frame remains of the dislocation notion which still conserves the invariance and continuity properties of the Burgers vectors. In the case of an optimum pattern, this pattern itself can constitute a periodic basis for secondary dislocations. The physical nature of these dislocations will have to be reconsidered since the relaxation in high-angle boundaries might be different from that in subgrain boundaries.

The extension of the 0-lattice theory to non-linear problems is discussed in Section 14.3, applied to examples where crystal dislocations end at a boundary or penetrate it.

In this manner the geometrical basis for arbitrary problems of crystalline interfaces is given. Future physical investigations may develop from here.

Suggestions for Problems

Problems of the type determining optimum phase boundaries usually go beyond the framework of exercises. It might be possible to construct such problems on simplified two-dimensional models which can be calculated by ordinary means (e.g. to find the optimum orientation of two lattices with slightly different lattice constants, determine the minimum geometrical parameter in order to calculate 0-lattice and cell walls for the optimum orientation).

Many exercises can be found for periodic patterns from Table 13.3/T1. The different periodic patterns can be constructed and checked on the moiré models. Also, the *DSC*-lattice and the secondary dislocation networks for slight deviations from the optimum orientation can be calculated.

Non-linear problems can hardly be solved without a computer.

Appendix

A1. Matrix Calculation Procedures

Since the 0-lattice theory is based on linear algebra, i.e. the matrix calculation, we present below the procedures involved in this technique. It is not our aim to develop and prove the mathematics, but rather to show how it is done. The proofs can be found in textbooks. A useful practical introduction is given by Wayman (1964).

The matrix calculation can be formally presented in an elegant manner, but it is often tedious for the practical calculation (e.g. by means of a desk calculator). On the other hand it is suitable for programming in computers, so that, without too much difficulty, problems can be treated which would be unthinkable on working by hand.

We restrict ourselves to the three- and the two-dimensional space. Unless otherwise mentioned, the three-dimensional space is assumed.

1. Coordinate Systems

All quantitative indications of positions are referred to a coordinate system, which is determined by an origin and three linearly independent basic vectors $\boldsymbol{u}_1, \boldsymbol{u}_2, \boldsymbol{u}_3$, such that every point in the space is determined by the position vector $\boldsymbol{x}$:

$$\boldsymbol{x} = x^1 \boldsymbol{u}_1 + x^2 \boldsymbol{u}_2 + x^3 \boldsymbol{u}_3. \qquad \text{(A1-1)}$$

The vectors are indicated by small, bold face letters, the coordinates in ordinary small letters. As is done in tensor calculations, we distinguish between lower and upper indices and apply Einstein's summation convention

$$x^i \boldsymbol{u}_i := \sum_{i=1}^{3} x^i \boldsymbol{u}_i. \qquad \text{(A1-2)}$$

If, in a term or a product, the same index appears above and below, it represents the summation over this index. The meaning of the upper and the lower indices will be discussed below.

In principle, any arbitrary coordinate system may be chosen. In our case, however, two types are especially useful:

a) The first is the orthogonal coordinate system with three mutually perpendicular basic vectors of the same unit length (of 1 Å, or the length of the cube edge in a cubic crystal system). The unit vectors are arranged in a right-handed sequence ($[\boldsymbol{u}_1 \times \boldsymbol{u}_2] = \boldsymbol{u}_3$). To facilitate plotting on graph paper it is advantageous to express the results in orthogonal coordinates.

b) The second is the crystal coordinate system with the unit vectors of the crystal lattice as basic vectors. The advantage is here that the crystal period is expressed as a "period" of integral numbers, so that the coordinates directly represent the position of the point with respect to the crystal unit cell.

Every crystal lattice is coupled with its "reciprocal lattice", which is often used in X-ray- and electron diffraction and also in solid state physics (Brillouin zones). The unit vectors of the reciprocal lattice are

$$\boldsymbol{u}^1 = \frac{[\boldsymbol{u}_2 \times \boldsymbol{u}_3]}{V} \tag{A 1-3}$$

with

$$V = (\boldsymbol{u}_1 [\boldsymbol{u}_2 \times \boldsymbol{u}_3]) \tag{A 1-4}$$

the volume of the crystal unit cell. The other unit vectors $\boldsymbol{u}^2$ and $\boldsymbol{u}^3$ follow by cyclic permutation.

The following orthogonality relation is fulfilled:

$$(\boldsymbol{u}_i \, \boldsymbol{u}^k) = \delta_i^k \tag{A 1-5}$$

$$\delta_i^k = \begin{cases} 1 & i = k \\ 0 & i \neq k \end{cases}. \tag{A 1-6}$$

A vector $\boldsymbol{x}$, referred to the reciprocal basis is:

$$\boldsymbol{x} = x_i \, \boldsymbol{u}^i. \tag{A 1-7}$$

In the *unit vectors* the lower indices $\boldsymbol{u}_i$ denote unit vectors of the crystal lattice, the upper indices $\boldsymbol{u}^i$ the unit vectors of the reciprocal lattice. In the coordinates, however, the upper indices x^i indicate that the vector is referred to the crystal coordinate system, and the lower indices x_i refer to the reciprocal coordinate system. x^i are often called the contravariant and x_i the covariant components of a vector. We shall not use this highly confusing notation here.

In the orthogonal coordinate system, the "crystal" system and the reciprocal coordinate system can be considered as identical (with the exception of the unit length Å and Å^{-1}). Hence there is no need to distinguish between lower and upper indices.

2. The Transition from the Orthogonal to the Crystal Coordinate System

A general crystal system is determined by the lattice constants $a, b, c, \alpha, \beta, \gamma$, where a, b, and c are the lengths of the unit vectors (in Å) and α, β, γ the angles between the unit vectors.

$$\alpha = \sphericalangle\, (\boldsymbol{b}, \boldsymbol{c}) \tag{A 1-8}$$

and correspondingly for β and γ (Section 2.2).

Starting from an orthogonal coordinate system (x, y, z), we have two degrees of freedom in the choice of the orientation of the crystal coordinate system within the orthogonal system. We may, for example, place the $\boldsymbol{c}$-axis on the z-axis (as is conventional in mineralogy) and the $\boldsymbol{a}$-axis in the xz-plane. [The notion (x, y, z) is used synonymously with $(\boldsymbol{u}_1\, \boldsymbol{u}_2\, \boldsymbol{u}_3)$ of the orthogonal coordinate system.] Now, we express the *new* crystal unit vectors $\boldsymbol{u}_1'\, \boldsymbol{u}_2'\, \boldsymbol{u}_3'$ (i.e. $\boldsymbol{a}$, $\boldsymbol{b}$, $\boldsymbol{c}$) in terms of the *old* orthogonal coordinates:

$$\begin{aligned} \boldsymbol{u}_1' &= \boldsymbol{u}_1\, s_1^1 + \boldsymbol{u}_2\, s_1^2 + \boldsymbol{u}_3\, s_1^3 \\ \boldsymbol{u}_2' &= \boldsymbol{u}_1\, s_2^1 + \boldsymbol{u}_2\, s_2^2 + \boldsymbol{u}_3\, s_2^3 \\ \boldsymbol{u}_3' &= \boldsymbol{u}_1\, s_3^1 + \boldsymbol{u}_2\, s_3^2 + \boldsymbol{u}_3\, s_3^3. \end{aligned} \tag{A 1-9}$$

Written in abbreviated form:

$$\boldsymbol{u}_i' = \boldsymbol{u}_k\, s_i^k \tag{A 1-10}$$

or in matrix notation

$$(\boldsymbol{u}_1'\, \boldsymbol{u}_2'\, \boldsymbol{u}_3') = (\boldsymbol{u}_1\, \boldsymbol{u}_2\, \boldsymbol{u}_3) \begin{pmatrix} s_1^1 & s_2^1 & s_3^1 \\ s_1^2 & s_2^2 & s_3^2 \\ s_1^3 & s_2^3 & s_3^3 \end{pmatrix} \tag{A 1-11}$$

and in abbreviated matrix notation:

$$\boldsymbol{w}'^T = \boldsymbol{w}^T \boldsymbol{S} \tag{A 1-12}$$

($|^T$ means transposed = row vector).

In this special case, $\boldsymbol{w}$ is a sort of a "super vector", the components of which are the unit vectors. Here we deal with pure vector calculation and do not yet decompose the vector into coordinates.

The general convention for ***matrix multiplication*** is:

$$\boldsymbol{C} = \boldsymbol{A}\boldsymbol{B}. \tag{A 1-13}$$

In index notation:

$$c_k^i = a_\alpha^i\, b_k^\alpha. \tag{A 1-14}$$

Sometimes it is convenient to denote the summing indices by Greek letters as in Eq. (A1-14). For example the element:

$$c_2^3 = a_1^3\, b_2^1 + a_2^3\, b_2^2 + a_3^3\, b_2^3 \tag{A 1-15}$$

i.e.

Eq. (A1-16)

$$\begin{pmatrix} \cdot & \cdot & \cdot \\ \cdot & \cdot & \cdot \\ \cdot & c_2^3 & \cdot \end{pmatrix} = \begin{pmatrix} \cdot & \cdot & \cdot \\ \cdot & \cdot & \cdot \\ a_1^3 & a_2^3 & a_3^3 \end{pmatrix} \begin{pmatrix} \cdot & b_2^1 & \cdot \\ \cdot & b_2^2 & \cdot \\ \cdot & b_2^3 & \cdot \end{pmatrix} \tag{A1-16}$$

This means that the scalar product of the *row- or line*-vector of the *first* matrix and the *column* vector of the *second* matrix is the element at the crossing point in the resultant matrix. The matrix product is usually not commutative ($\boldsymbol{AB} \neq \boldsymbol{BA}$). If the first matrix consists of only one row (row vector), then the resulting matrix consists also of one row. Correspondingly, if the second matrix is a column vector, the result is also a column vector.

Remark

We combine here matrix- and tensor notation. In the matrix notation the position of the elements in the matrix is determined by the sequence of the indices and the matrix multiplication is non-commutative. In the tensor notation, however, the coefficients form a set, where every element is identified by its indices but has no fixed position in a matrix. Hence, in the tensor notation the indices can be placed on top of each others and the multiplication of the coefficients is communtative.

In the combined matrix-tensor notation Eq. (Al-14) should be written:

$$c^i{}_k = a^i{}_\alpha \, b^\alpha{}_k \,. \tag{A1-14a}$$

Then the transpose of $\boldsymbol{C}$: $\boldsymbol{C}^T = \boldsymbol{B}^T \boldsymbol{A}^T$ is:

$$c_k{}^i = b_k{}^\alpha \, a_\alpha{}^i \,. \tag{A1-14b}$$

In tensor notation

$$c_k^i = b_k^\alpha \, a_\alpha^i \tag{A1-14c}$$

is the same as Eq. (Al-14) as only coefficients are multiplied and added according to the summation convention and not rows an columns of matrices. In this book, where the indices are placed vertically above each others the upper index has to be considered as the first.

Our task now is to determine the coefficients of the $\boldsymbol{S}$-matrix [Eq. (A1-11)]. If the $\boldsymbol{c}$-axis is placed on the z-axis, then

$$s_3^3 = c \tag{A1-17}$$

and

$$s_3^1 = s_3^2 = 0 \tag{A1-18}$$

and, if the $\boldsymbol{a}$-axis is placed in the (xz)-plane:

$$s_1^2 = 0. \tag{A 1-19}$$

The remainder of the coefficients are determined from Eq. (A 1-9) by the scalar products of the vector $\boldsymbol{u}_i'$ by using the orthogonality of the $\boldsymbol{u}_i$:

$$(\boldsymbol{u}_i\,\boldsymbol{u}_k) = \delta_{ik} \tag{A 1-20}$$

and the definition (A 1-8).

For example:

$$(\boldsymbol{u}_1'\,\boldsymbol{u}_3') = a \cdot c \cdot \cos\beta = s_1^3 \cdot s_3^3 = s_1^3 \cdot c \tag{A 1-21}$$

i.e.

$$s_1^3 = a \cdot \cos\beta. \tag{A 1-22}$$

Further equations are given by the scalar products of the vectors $\boldsymbol{u}_i'$ with themselves, e.g.

$$(\boldsymbol{u}_1'\,\boldsymbol{u}_1') = a^2 = (s_1^1)^2 + (s_1^3)^2 = (s_1^1)^2 + a^2\cos^2\beta \tag{A 1-23}$$

i.e.

$$(s_1^1)^2 = a^2(1 - \cos^2\beta) = a^2\sin^2\beta \tag{A 1-24}$$

and

$$s_1^1 = a \cdot \sin\beta. \tag{A 1-25}$$

In this way, the whole **S**-matrix is obtained. Due to the lack of space, we write the column vectors below, instead of beside one another.

$$\left.\begin{array}{ll}
\boldsymbol{a} & s_1^1 = a \cdot \sin\beta \\
 & s_1^2 = 0 \\
 & s_1^3 - a \cdot \cos\beta \\
\boldsymbol{b} & s_2^1 = (b/\sin\beta)(\cos\gamma - \cos\alpha\cos\beta) \\
 & s_2^2 = (b/\sin\beta)\,[2\cos\alpha\cos\beta\cos\gamma + \sin^2\alpha\sin^2\beta \\
 & \qquad - \cos^2\alpha\cos^2\beta - \cos^2\gamma]^{\frac{1}{2}} \\
 & s_2^3 = b \cdot \cos\alpha \\
\boldsymbol{c} & s_3^1 = 0 \\
 & s_3^2 = 0 \\
 & s_3^3 = c
\end{array}\right\} \tag{A 1-26}$$

The column vectors of the S-matrix are thus the unit vectors of the crystal coordinate system expressed in orthogonal coordinates.

The basic vectors of any crystal system can be expressed in this form by specifying the lattice constants. Summarizing: The S-matrix is determined by expressing the new *basic vectors* in terms of the old ones, without decomposing the vectors into coordinates.

3. Coordinate- and vector transformations

We assume now that both the coordinate systems: the orthogonal and the crystal system, are given, and consider an arbitrary vector $\boldsymbol{x}$ with the orthogonal coordinates ξ^i and the crystal coordinates ξ'^i

$$\boldsymbol{\xi} = \boldsymbol{u}_i \xi^i = \boldsymbol{u}'_i \xi'^i. \qquad \text{(A 1-27)}$$

By using the Greek letter ξ^i, ξ'^i we stress the fact that they represent the *same* vector in *different* coordinate systems. The transition between the coordinates is given by:

$$\boldsymbol{\xi} = \boldsymbol{S}\,\boldsymbol{\xi}'. \qquad \text{(A 1-28)}$$

Therefore, the $\boldsymbol{S}$-matrix transforms the crystal coordinates of a vector (e.g. $\boldsymbol{\xi}'^{(1)} = [1\,0\,0]$) into its orthogonal coordinates ($\boldsymbol{\xi}^{(1)} = a \cdot \sin\beta$, 0, $a \cdot \cos\beta$). We term this transition, in which the vector is fixed but the coordinate system changes, a *coordinate transformation.*

Eq. (A 1-28) takes another meaning when written in the form:

$$\boldsymbol{x}' = \boldsymbol{S}\boldsymbol{x}. \qquad \text{(A 1-29)}$$

Here we choose the orthogonal coordinate system and consider it as fixed, but now $\boldsymbol{x}$ and $\boldsymbol{x}'$ represent two different vectors. For example, the vector $[1\,0\,0]$ in the orthogonal system (x-edge of the unit cube) is transformed into the vector $\boldsymbol{a} = [a \cdot \sin\beta, 0, a \cdot \cos\beta]$ in the same coordinate system. The vector has changed its length from 1 Å to a Å and is rotated by $\pi/2 - \beta$ in the positive sense in the xz-plane. The complete orthogonal unit cube is transformed by $\boldsymbol{S}$ into the crystal unit cell. We call this transition between two *different* vectors a *vector transformation.*

The inverse coordinate transformation is given by

$$\boldsymbol{\xi}' = \boldsymbol{S}^{-1}\,\boldsymbol{\xi} \qquad \text{(A 1-30)}$$

and the inverse vector transformation by

$$\boldsymbol{x} = \boldsymbol{S}^{-1}\,\boldsymbol{x}'. \qquad \text{(A 1-31)}$$

[The calculation of the inverse matrix $\boldsymbol{S}^{-1}$ will be given in Eq. (A1-36), (A 1-37).]

It is important to know, in every case, whether one is dealing with a coordinate transformation or with a vector transformation.

4. The Determinant and the Inverse Matrix

The *determinant* of a matrix is the following sum of diagonal products:

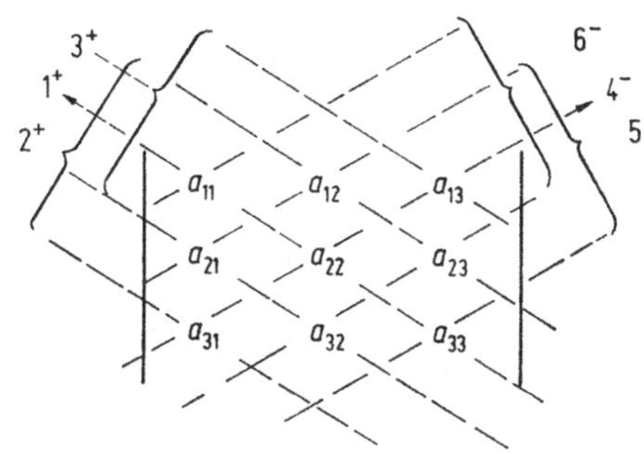

$$= \underbrace{a_{11}\,a_{22}\,a_{33}}_{(1)} + \underbrace{a_{12}\,a_{23}\,a_{31}}_{(2)} + \underbrace{a_{13}\,a_{21}\,a_{32}}_{(3)} - \underbrace{a_{13}\,a_{22}\,a_{31}}_{(4)} - \underbrace{a_{12}\,a_{21}\,a_{33}}_{(5)} - \underbrace{a_{11}\,a_{23}\,a_{32}}_{(6)} \tag{A1-32}$$

For two-row matrices, the determinant is:

Eq. (A1-33):

$$\overset{(+)\qquad(-)}{\begin{vmatrix} a_{11} & a_{12} \\ a_{21} & a_{22} \end{vmatrix}} = a_{11}\,a_{22} - a_{21}\,a_{12}. \tag{A1-33}$$

As we shall see, the determinant is a number, a ratio.

In order to calculate what is called the *subdeterminant* A_{ik} of an element a_{ik} of the general matrix $\boldsymbol{A}$, the vertical and horizontal rows containing the element a_{ik} are crossed out. The determinant of the remaining two-row matrix is calculated and multiplied by $(-1)^{i+k}$ in order to fix the sign, i.e.

Eq. (A1-34):

$$A_{23} = \begin{vmatrix} a_{11} & a_{12} & a_{13} \\ a_{21} & a_{22} & \boxed{a_{23}} \\ a_{31} & a_{32} & a_{33} \end{vmatrix} = (-1)^{2+3}(a_{11}\,a_{32} - a_{12}\,a_{31}). \tag{A1-34}$$

$$= a_{12}\,a_{31} - a_{11}\,a_{32}$$

The subdeterminant of a two-row matrix A_{21} is:

$$A_{21} = (-1)^{2+1} \cdot a_{12} = -a_{12}. \tag{A1-35}$$

An important concept is the *rank* of a matrix. The rank is the number of rows (columns respectively) of the largest subdeterminant (determinant resp.) with non-zero value. If for a three-row matrix the determinant $|\boldsymbol{A}| \neq 0$, then its rank is 3. If, however, $|\boldsymbol{A}| = 0$, and at least one two-row subdeterminant $A_{ik} \neq 0$, then $\operatorname{rank}(\boldsymbol{A}) = 2$, etc.

The elements of the *inverse matrix* are determined by:

$$a_{ik}^{(-1)} = \frac{A_{ki}}{|\boldsymbol{A}|}. \tag{A 1-36}$$

In order to calculate the element $a_{23}^{(-1)}$ of the inverse matrix, the subdeterminant of the element a_{32} (the mirror image across the main diagonal) is calculated and divided by the determinant of $\boldsymbol{A}$.

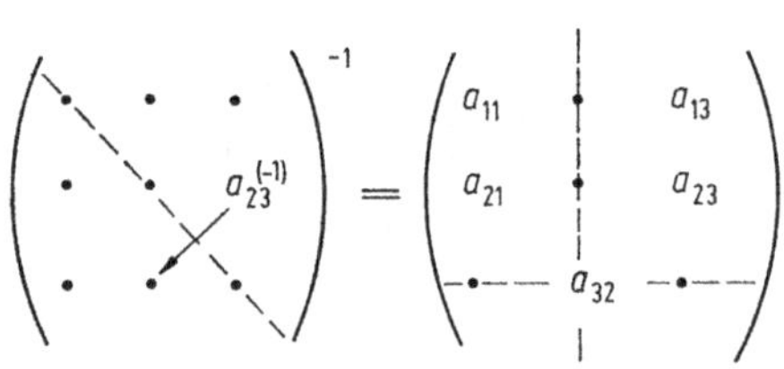

(A 1-37)

$$a_{23}^{(-1)} = \frac{A_{32}}{|\underline{A}|} = \frac{(-1)^{2+3}\,(a_{11}a_{23} - a_{13}a_{21})}{|\underline{A}|}$$

The definition of the inverse matrix implies that

$$\boldsymbol{A}\boldsymbol{A}^{-1} = \boldsymbol{A}^{-1}\boldsymbol{A} = \boldsymbol{I} \tag{A 1-38}$$

with $I = \delta_k^i =$ identity.

$$\boldsymbol{I} = \begin{pmatrix} 1 & 0 & 0 \\ 0 & 1 & 0 \\ 0 & 0 & 1 \end{pmatrix}. \tag{A 1-39}$$

As a check, it is recommended to calculate the product (A 1-38) every time an inverse matrix has been determined.

An inverse matrix only exists if the determinant $|\boldsymbol{A}| \neq 0$, as shown by Eq. (A 1-36).

By means of the inverse matrix, for example, Eq. (A 1-29) may be solved for $\boldsymbol{x}$:

$$\boldsymbol{x}' = \boldsymbol{S}\boldsymbol{x}. \tag{A 1-40}$$

Both sides are multiplied from the left by $\boldsymbol{S}^{-1}$:

$$\boldsymbol{S}^{-1}\boldsymbol{x}' = \boldsymbol{S}^{-1}\boldsymbol{S}\boldsymbol{x} = \boldsymbol{I}\boldsymbol{x} = \boldsymbol{x} \tag{A 1-41}$$

which is the same as Eq. (A 1-31).

Another application of the inverse matrix is the determination of the basic vectors of the reciprocal lattice $\boldsymbol{u}'^i$, expressed in orthogonal coordinates. On the one hand the $\boldsymbol{u}'^i$ and $\boldsymbol{u}'_i$ are related by Eq. (A 1-5):

$$\boldsymbol{u}'^i \boldsymbol{u}'_k = \delta^i_k \tag{A 1-42}$$

($\boldsymbol{u}'_i$ = general unit vectors, $\boldsymbol{u}_i$ = orthogonal unit vectors). On the other hand:

$$\boldsymbol{S}^{-1}\boldsymbol{S} = \boldsymbol{I}. \tag{A 1-43}$$

According to the matrix multiplication convention a row vector of the first matrix is multiplied as a scalar product with a column vector of the second matrix. Here, the column vectors of $\boldsymbol{S}$ are the vectors $\boldsymbol{u}'_k$, hence the row vectors of $\boldsymbol{S}^{-1}$ must be the $\boldsymbol{u}'^i$. Thus, the vector products of Eq. (A 1-3) can be replaced by the matrix inversion in order to determine the vectors $\boldsymbol{u}'^i$.

5. Metrics

The *metric tensor* is defined in the following way:

$$\boldsymbol{G} = g_{ik} = (\boldsymbol{u}'_i \boldsymbol{u}'_k). \tag{A 1-44}$$

The terms of $\boldsymbol{G}$ are thus the scalar products of the basic vectors. Since the scalar products are commutative, the metric tensor is always *symmetrical.*

$$g_{ik} = g_{ki}. \tag{A 1-45}$$

In the *orthogonal system,* $\boldsymbol{G}$ is, according to Eq. (A 1-20), given by

$$\boldsymbol{G} = g_{ik} = (\boldsymbol{u}_i \boldsymbol{u}_k) = \delta_{ik} = \boldsymbol{I}. \tag{A 1-46}$$

$\boldsymbol{G}$ can be determined from the $\boldsymbol{S}$-matrix by:

$$\boldsymbol{G} = \boldsymbol{S}^T \boldsymbol{S} \tag{A 1-47}$$

with $\boldsymbol{S}^T$ the transposed matrix of $\boldsymbol{S}$. (The transposed matrix is the mirror image with respect to the main diagonal.)

$$(s^i{}_k)^T = s_k{}^i. \tag{A 1-48}$$

The metric tensor is a characteristic of the coordinate system. It helps to determine metric properties such as absolute length and angles in general coordinate systems. The *scalar product* of two vectors is:

$$|\boldsymbol{x}| \cdot |\boldsymbol{y}| \cos\phi = \boldsymbol{x}^T \boldsymbol{G} \boldsymbol{y} = x^i g_{ik} y^k \tag{A 1-49}$$

and, as a special case, the square of the absolute length is

$$|\boldsymbol{x}|^2 = \boldsymbol{x}^T \boldsymbol{G} \boldsymbol{x} = x^i g_{ik} x^k \tag{A 1-50}$$

and the angle ϕ between the vectors $\boldsymbol{x}$ and $\boldsymbol{y}$ is given by:

$$\cos\phi = \frac{\boldsymbol{x}^T \boldsymbol{G} \boldsymbol{y}}{|\boldsymbol{x}|\,|\boldsymbol{y}|} = \frac{x^i g_{ik} y^k}{[(x^i g_{ik} x^k)(y^i g_{ik} y^k)]^{\frac{1}{2}}}. \tag{A 1-51}$$

Another way would be to transform the vectors $\boldsymbol{x}$ and $\boldsymbol{y}$ first to the orthogonal coordinate system and then to determine the length and angle by the standard orthogonal scalar products $(x_i\, y_i)$.

The metric tensor is also required for the coordinate transformation from the crystal coordinate system to the reciprocal coordinate system.

$$\xi_i = g_{ik}\,\xi^k. \qquad \text{(A 1-52)}$$

The metric tensor of the reciprocal coordinate system is:

$$g^{ik} = \boldsymbol{G}^{-1} \qquad \text{(A 1-53)}$$

and thus the inverse coordinate transformation is

$$\xi^i = g^{ik}\,\xi_k. \qquad \text{(A 1-54)}$$

(In the orthogonal coordinate system $\boldsymbol{G}=\boldsymbol{I}$, therefore $\xi^i = \xi_i$. Hence, disregarding the dimension Å or Å^{-1}, the distinction between upper and lower indices is meaningless in this case.) The scalar product can also be written as:

$$(\boldsymbol{x}\boldsymbol{y}) = (x^i\, y_i) \qquad \text{(A 1-55)}$$

$$(\boldsymbol{x}\boldsymbol{x}) = |\boldsymbol{x}|^2 = (x^i\, x_i) \qquad \text{(A 1-56)}$$

and

$$\cos\phi = \frac{x^i\, y_i}{[(x^i\, x_i)\,(y^i\, y_i)]^{\frac{1}{2}}}. \qquad \text{(A 1-57)}$$

Summarizing the *coordinate* transformations, we obtain:

$$\begin{array}{ccccc} \textit{orthogonal} & & \textit{crystal} & & \textit{reciprocal} \\ \xi^{(\mathrm{orth})} & \underset{\boldsymbol{S}}{\overset{\boldsymbol{S}^{-1}}{\rightleftarrows}} & \xi^{(c)} & \underset{\boldsymbol{G}^{-1}}{\overset{\boldsymbol{G}}{\rightleftarrows}} & \xi^{(r)}. \end{array}$$

6. General Linear Transformations

A linear transformation $\boldsymbol{A}$ describes an affine deformation (leaving parallel lines parallel) of the whole space with all its vectors. Such transformations are, for example, expansion, rotation, and shear. A transformation is said to be *homogeneous* if the point at the origin remains fixed (no constant terms in the equations). If the transformation contains a translation, it is termed *nonhomogeneous*. We shall only discuss homogeneous transformations.

If by the transformation $\boldsymbol{A}$ every vector $\boldsymbol{x}$ is related to a distinct vector $\boldsymbol{x}'$ and vice versa

$$\boldsymbol{x}' = \boldsymbol{A}\boldsymbol{x} \qquad x'^i = a^i_k\, x^k \qquad \text{(A 1-58)}$$

the transformation is characterized as *regular* or *non degenerate*. In this case the determinant $|\boldsymbol{A}| \neq 0$ and the inverse matrix $\boldsymbol{A}^{-1}$ can be calculated

[Eq. (A 1-36)]. (In a non regular case, for example, a three-dimensional space would be imaged onto a two-dimensional one.)

The coefficients of the matrix $\boldsymbol{A}$ depend, of course, on the coordinate system in which the transformation is represented. Many transformations acquire an especially convenient form in the orthogonal coordinate system.

Examples in orthogonal coordinates are:

a) An *expansion* (or contraction) by the factor a in the direction of the x-axis, by b in the y-axis and by c in the z-axis is described by:

$$\text{Expansion } \boldsymbol{E} = \begin{pmatrix} a & 0 & 0 \\ 0 & b & 0 \\ 0 & 0 & c \end{pmatrix}. \tag{A 1-59}$$

b) A convenient representation of a *rotation* in the right-handed screw sense by the angle θ around an axis given by the unit vector $[c_1, c_2, c_3]$ (c_i = direction cosines) is the following:

$$\left.\begin{aligned} r_{ik} &= \delta_{ik} \cos\theta + c_i c_k (1 - \cos\theta) + c_{ik} \sin\theta \\ c_{ik} &= \begin{pmatrix} 0 & -c_3 & c_2 \\ c_3 & 0 & -c_1 \\ -c_2 & c_1 & 0 \end{pmatrix}. \end{aligned}\right\} \tag{A 1-60}$$

A special advantage of the rotations (orthogonal transformation) in the orthogonal coordinate system is that:

$$\boldsymbol{R}^{-1} = \boldsymbol{R}^T. \tag{A 1-61}$$

In order to determine the inverse of $\boldsymbol{R}$, the mirror image of the matrix has to be written with respect to the main diagonal.

An important application is the rotation of an arbitrary vector $\boldsymbol{v}$ into a special position, e.g. into the z-axis. Such a vector can be the normal to the b-subspace or the invariant eigenvector. [Standard form Eq. (12.9-22).] The given vector $\boldsymbol{v}$ is assumed to be normalized ($|\boldsymbol{v}| = 1$). Its components are $[v_1, v_2, v_3]$. At the same time they are the direction cosines, i.e. for example, $v_3 = \cos \measuredangle(\boldsymbol{v}, \boldsymbol{u}_3) = \cos\theta$. In order to rotate the vector $\boldsymbol{v}$ into the z-axis, we have to rotate it around an axis which is perpendicular to both, $\boldsymbol{v}$ and $\boldsymbol{u}_3$. This is obtained from the vector product $[\boldsymbol{v} \times \boldsymbol{u}_3]$:

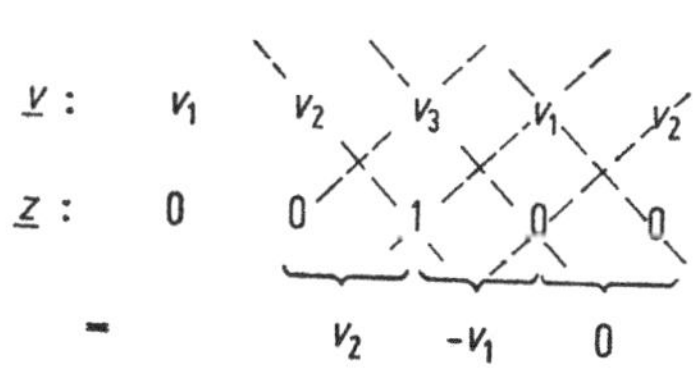

(A 1-62)

This new vector is also normalized:

$$c=\left\{\frac{v_2}{(v_1^2+v_2^2)^{\frac{1}{2}}},\quad \frac{-v_1}{(v_1^2+v_2^2)^{\frac{1}{2}}},\ 0\right\}. \tag{A 1-63}$$

In this way, the vector $\boldsymbol{c}$ required for the rotation Eq. (A 1-60) is determined. Further: $\cos\theta=v_3$ and $\sin\theta=(1-v_3^2)^{\frac{1}{2}}$, so that now the data needed for the formulation of the rotation matrix are all known.

c) A pure *shear* in the direction of the y-axis, with the (xy)-plane invariant, is described by:

$$\text{Shear } \boldsymbol{C}=\begin{pmatrix}1 & 0 & 0\\ 0 & 1 & \sigma\\ 0 & 0 & 1\end{pmatrix}. \tag{A 1-64}$$

The *determinant* of a matrix $\boldsymbol{A}$ of a linear transformation is the ratio between the new and the old unit volumes. $\boldsymbol{A}$ is here understood as the transformation of the basic vectors

$$\boldsymbol{u}_i'=\boldsymbol{u}_k\, a_i^k. \tag{A 1-65}$$

Then

$$|\boldsymbol{A}|=\frac{V'}{V}. \tag{A 1-66}$$

A transformation for which $|\boldsymbol{A}|=1$ leaves the unit volume unchanged. It is called a *unimodular* transformation (e.g. pure shear). If, in addition, the distances between all the points stay unchanged the transformation is termed *orthogonal* (general rotation + mirror image).

7. Combination of Transformations

Several consecutive transformations can be applied to a vector, where the later transformation acts on the result of the earlier transformation. For example, a space can be sheared in the direction of the y-axis [transformation $\boldsymbol{C}$ Eq. (A 1-64)] and then be expanded by $1+\varepsilon_z$ in the direction of the z-axis [Eq. (A 1-59), $a=b=1$, $c=1+\varepsilon_z$]. The result is given by:

$$\boldsymbol{x}''=\boldsymbol{E}\boldsymbol{C}\boldsymbol{x}. \tag{A 1-67}$$

Arbitrary transformations may be constructed in this way. The convention of the matrix calculation is such that the *sequence of the operations* goes from *right to left*, opposite to the sense of writing.

8. Transformation of Transformations

It may occur, for example, that a pure shear has to be represented in such a way that the orientation of the shear direction and the shear plane are given in space. We may start from a special representation of the type of Eq. (A 1-64), and then rotate the whole transformation.

We have to decide here whether we leave the coordinate system fixed and transform the vector space or whether we rotate the coordinate system. This is essentially a question of the sign. It is important for the representation whether, for example, the vector space or the coordinate system is rotated by $+30°$ around the positive z-axis. The analytical representation, however, is the same for a rotation of the vector space by $+30°$ or of the coordinate system by $-30°$.

Let us consider for the moment the rotation of the vector space. The rotation $\boldsymbol{R}$ shall be defined in the same way in which $\boldsymbol{S}$ is defined [Eq. (A 1-11), (A 1-60), (A 1-61)] such that, in the *fixed* coordinate system, the new vector follows from the old by:

$$\boldsymbol{x}' = \boldsymbol{R}\boldsymbol{x}. \tag{A 1-68}$$

We now apply this rotation on the general transformation $\boldsymbol{A}$ (e.g. shear). The *rotated transformation* together with its characteristic elements is then given by (vector transformation):

$$\boldsymbol{A}' = \boldsymbol{R}\boldsymbol{A}\boldsymbol{R}^{-1}. \tag{A 1-69}$$

If, however, the *coordinate system is rotated* by $\boldsymbol{R}$, then the transformation $\boldsymbol{A}''$ represented in the new coordinate system (coordinate transformation) is:

$$\boldsymbol{A}'' = \boldsymbol{R}^{-1}\boldsymbol{A}\boldsymbol{R}. \tag{A 1-70}$$

It might, for example, happen that the simple shear in the special orientation $\boldsymbol{C}$ in a fixed orthogonal coordinate system has to be rotated into another orientation (vector transformation):

$$\boldsymbol{C}' = \boldsymbol{R}\boldsymbol{C}\boldsymbol{R}^{-1} \tag{A 1-71}$$

then the new general shear has to be represented in terms of crystal coordinates [coordinate transformation, Eq. (A 1-29)]

$$\boldsymbol{C}^{(c)} = \boldsymbol{S}^{-1}\boldsymbol{C}'\boldsymbol{S} = \boldsymbol{S}^{-1}\boldsymbol{R}\boldsymbol{C}\boldsymbol{R}^{-1}\boldsymbol{S}. \tag{A 1-72}$$

The same transformation now represented in the reciprocal coordinate system [$\boldsymbol{\xi}' = \boldsymbol{G}\boldsymbol{\xi} \to \boldsymbol{x} = \boldsymbol{G}\boldsymbol{x}' \to \boldsymbol{x}' = \boldsymbol{G}^{-1}\boldsymbol{x}$ i.e. $\boldsymbol{G}^{-1} \approx \boldsymbol{R}$ in Eq. (A 1-70)] is:

$$\boldsymbol{C}^{(r)} = \boldsymbol{G}\boldsymbol{C}^{(c)}\boldsymbol{G}^{-1} = \boldsymbol{G}\boldsymbol{S}^{-1}\boldsymbol{R}\boldsymbol{C}\boldsymbol{R}^{-1}\boldsymbol{S}\boldsymbol{G}^{-1}. \tag{A 1-73}$$

In this way, arbitrary transformations in arbitrary coordinate systems can be constructed. (A control of tedious matrix calculations is given by the determinants as $|\boldsymbol{ABC}| = |\boldsymbol{A}| \cdot |\boldsymbol{B}| \cdot |\boldsymbol{C}|$ and $|\boldsymbol{A}^{-1}| = 1/|\boldsymbol{A}|$.)

We have to distinguish clearly between the combination of transformations and the transformation of transformations. When a shear in the y-direction [Eq. (A 1-64)] is combined with a rotation by $\theta = 90°$ around the positive z-axis, the vector space is sheared beforehand and then rotated such that, as a result of the combined transformation $\boldsymbol{RC}$, the following vectors are related: $\boldsymbol{x}[100] \to \boldsymbol{x}'[010]$ and $\boldsymbol{x}[001] \to \boldsymbol{x}'[-\sigma, 0, 1]$. If, however, the shearing is *transformed* by a rotation ($\boldsymbol{RCR}^{-1}$) then the following vectors are related: $\boldsymbol{x}[100] \to \boldsymbol{x}'[100]$ and $\boldsymbol{x}[001] \to \boldsymbol{x}'[-\sigma, 0, 1]$. The vector space is only sheared but in a new direction. This example may be calculated as an exercise. In any case it is to be recommended in doubtful cases to calculate model examples.

9. Addition of Transformations

The addition of two transformations $\boldsymbol{A}+\boldsymbol{B}$ means the addition of transformed vectors $\boldsymbol{x}'$ originating from the same vector $\boldsymbol{x}$

$$\boldsymbol{C}\boldsymbol{x} = (\boldsymbol{A}+\boldsymbol{B})\,\boldsymbol{x} = \boldsymbol{A}\boldsymbol{x} + \boldsymbol{B}\boldsymbol{x}. \tag{A1-74}$$

The matrices are added element by element

$$c_{ik} = a_{ik} + b_{ik}. \tag{A1-75}$$

10. Invariant Elements of a Transformation

By a non-degenerate linear transformation, the vector space as a whole is transformed into itself. In the case of a homogeneous transformation the origin stays unchanged $\boldsymbol{x}[000] \to \boldsymbol{x}'[000]$. We may ask whether further unchanged elements exist. Vectors with invariant orientation (while their length may change), are called *eigenvectors*. They are determined by the following condition:

$$\boldsymbol{A}\boldsymbol{x} = \lambda\boldsymbol{x} = \boldsymbol{x}'. \tag{A1-76}$$

The numerical parameter λ is called the *eigenvalue*. Its meaning is the ratio of the length of the eigenvector after and before the transformation

$$\lambda = \frac{|\boldsymbol{x}'|}{|\boldsymbol{x}|}. \tag{A1-77}$$

The equation for the eigenvectors is:

$$(\boldsymbol{A} - \lambda\boldsymbol{I})\,\boldsymbol{x} = \boldsymbol{0}. \tag{A1-78}$$

where $\boldsymbol{0}$ means the zero vector $[000]$. This equation has non-trivial solutions only when:

$$|\boldsymbol{A} - \lambda\boldsymbol{I}| = 0. \tag{A1-79}$$

Eq. (A1-79) is an equation of the third degree for λ. The (real) solutions λ are introduced into Eq. (A1-78), from which the eigenvectors $\boldsymbol{x}$ can be calculated.

If we refer to the end vector $\boldsymbol{x}'$ instead of to the starting vector $\boldsymbol{x}$, and denote this end vector by $\boldsymbol{x}^{(0)}$, Eq. (A1-78) becomes:

$$(\boldsymbol{I} - \lambda\boldsymbol{A}^{-1})\,\boldsymbol{x}^{(0)} = \boldsymbol{0}. \tag{A1-80}$$

The eigenvalues λ are determined in an analogous manner to Eq. (A 1-79). If we abbreviate

$$(\boldsymbol{I}-\lambda \boldsymbol{A}^{-1})\, \boldsymbol{x}^{(0)} := \boldsymbol{T}\boldsymbol{x}^{(0)} \qquad \text{(A 1-81)}$$

this equation becomes, in index notation:

$$t^i_k\, x^{(0)\,k} = 0. \qquad \text{(A 1-82)}$$

The eigenvectors can be determined by the procedure given in Section 12.9 [Eq. (12.9-9)] etc.

For $\lambda=1$, not only the direction but also the length of the eigenvector is conserved. In this case, it is called an *invariant* eigenvector. If we compare (A 1-80) with the basic equation of the 0-lattice [Eq. (12.3-6)]:

$$(\boldsymbol{I}-\boldsymbol{A}^{-1})\, \boldsymbol{x}^{(0)} = \boldsymbol{b}^{(L)} \qquad \text{(A 1-83)}$$

we see that the 0-element passing through the origin ($\boldsymbol{b}^{(L)}=\boldsymbol{0}$) contains the invariant eigenvector.

This collection of procedures of matrix calculation cannot naturally be complete. The best way to get acquainted with the technique is to calculate examples with directly evident results.

A 2. Moiré-Models

Four pairs of moiré models and a scale $\boldsymbol{M}$ are joined to the book. The model 1 corresponds to a {1 1 0}-face in the body-centred cubic (bcc) structure. The point arrangement is approximately hexagonal. The length of the cube edge is chosen as 2 mm. Model 2 represents a {1 1 2}-face and model 3 a {1 0 0}-face in the same structure. Model 4 consists of parallel lines also with 2 mm spacing. Fig. A 2/1 shows the dimensional situation.

Every model consists of a sheet $\boldsymbol{a}$ and a sheet $\boldsymbol{b}$. Sheet $\boldsymbol{a}$ may be placed below the scale and its position fixed by means of the straight scales. The zero position on all four linear scales corresponds to the exact centering. Sheet $\boldsymbol{b}$ can be placed on top of the scales, and its position is determined with respect to rotation. As moiré figures are extremely sensitive to changes in relative position, a high precision is needed, especially if calculated patterns are to be compared quantitatively with the models.

It was shown in Section 12.10 that, with respect to translation a shift $\boldsymbol{d}^{(2)}$ is equivalent to one by $\boldsymbol{d}^{(1)}\,(=\boldsymbol{d}^{(b)})$, where $\boldsymbol{d}^{(1)}=-\boldsymbol{A}^{-1}\,\boldsymbol{d}^{(2)}$. Instead of executing the rotation and the translation with sheet $\boldsymbol{b}$, the translation can be done with sheet $\boldsymbol{a}$ in coordinates of crystal 1 and the rotation alone with sheet $\boldsymbol{b}$. The angle of rotation is best fixed with all four marks simultaneously on the circular segments. The coordinate

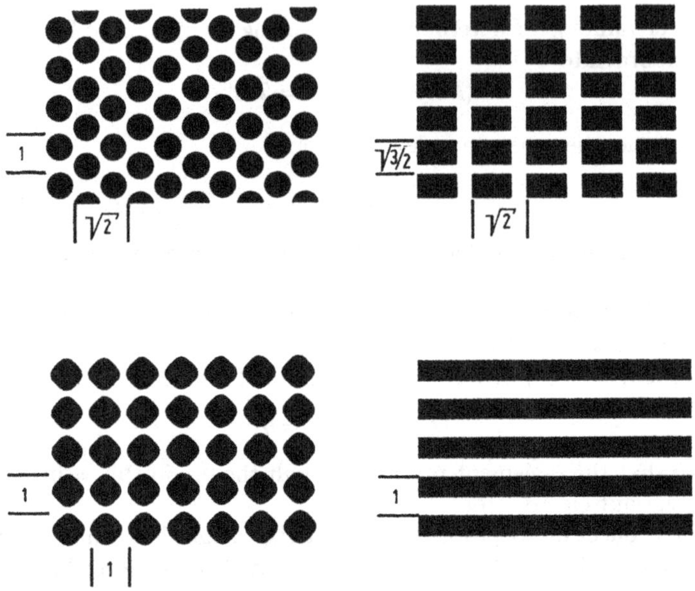

Fig. A2/1. Dimensions of the moiré models (the unit corresponds to 2 millimeters)

origin in this case is the centre of the scale sheet, i.e. also the centre of sheet ***b***.

The upper left-hand corner of the given lattices are cut such that the sheets can always be placed on top of each other in the same orientation. In this way eventual irregularities in the drawing or distortions due to the photographic reproduction may cancel out optimally.

In order to add an ***expansion*** to the rotation, the aspect of perspective can be applied by keeping the sheets apart when viewing them. The movement rule [Eq. (5.4-1)] for edge dislocations can be checked in this way with model 4. Expansion combined with rotation and translation can also be controlled quantitatively by mounting the sheets on glass plates and photographing them above a luminous table by means of a reflex camera.

The models are transparent in view of their possible use as demonstration material in lectures, where they can be projected. Here a viewgraph is recommended, in which the sheets ***M*** and ***a*** can be placed under- and the sheet ***b*** on top of the transparent plastic foil used for writing. In this way the displacements of sheet ***b*** do not affect the lower sheets. An epidiascope should not be used for projection, as the models are likely to be burnt.

Many calculations can be controlled by means of the models, especially in connection with Table 13.3/T1. Also, the behaviour of the pattern on translation of one lattice with respect to the other can be studied.

Point lattices have been treated exclusively until now. In order to calculate, for example, the rotation of a line lattice (moiré model No. 4) the line lattice can first be treated as a point lattice, which leads to Eq. (12.3-7) with $(\boldsymbol{I}-\boldsymbol{A}^{-1})^{-1}$ in the form of Eq. (12.3-12). Then, in the final result, the $b_1^{(L)}$-component is assumed to be continuous while $b_2^{(L)}$ stays discrete; i.e. $b_1^{(L)}=\xi$ and $b_2^{(L)}=n$. Afterwards, ξ is eliminated, from which follows the equation of the moiré lines:

$$x_2^{(0)}=-x_1^{(0)}\cot\mathrm{an}(\theta/2)+n/\left(2\sin^2(\theta/2)\right). \tag{A2-1}$$

A wide range of moiré problems can be solved by the general non-linear method given in Section 14.3; for example, the superposition of circular waves. The calculation shall be briefly described here:

Origins: wave 1 $(d/2, 0)$
wave 2 $(-d/2, 0)$

$$x^{(1)}(z+b^{(L)})=\left(x-(d/2)\right)^2+y^2-(z+n)^2=0 \tag{A2-2}$$

$$x^{(2)}(z)=\left(x+(d/2)\right)^2+y^2-z^2=0. \tag{A2-3}$$

From

$$x^{(2)}(z)-x^{(1)}(z+b^{(L)})=0 \tag{A2-4}$$

follows

$$z=-\frac{xd}{n}-\frac{n}{2} \tag{A2-5}$$

in Eq. (A2-2):

$$\frac{x^2}{\frac{n^2}{4}}-\frac{y^2}{\frac{d^2-n^2}{4}}=1. \tag{A2-6}$$

For $|n|<|d|$, the equation represents hyperbola, and for $|n|>|d|$ ellipses (which arise if one wave is moving outwards and the other inwards). (For further details on moiré patterns, see G. Oster.)

A3. The Direct Observation of Crystal Defects

We give here a few indications on the literature dealing with the direct observation of crystal defects. The method which shows directly the atomic arrangement is field ion microscopy, which was developed by E. W. Müller. It is described among others by D. G. Brandon (1966). The method which is mostly used today for the study of dislocations and stacking faults is transmission electron microscopy which, based on work of R. D. Heidenreich and also R. Castaing, was developed by the group P. B. Hirsch, M. J. Whelan, R. W. Horne and independently by the author. Books are indicated in the bibliography. X-ray topography also permits the direct imaging of crystal defects in an analogous manner to transmission electron microscopy, on much thicker specimens with, however, lower magnification (see e.g. Brandon, 1966). Somewhat less direct methods are on the one hand etch pits, which form, on etching a crystal, at locations where dislocations end at the crystal surface, and on the other hand the surface decoration (developed by Basset and further by Bethge). Here, gold is lightly evaporated on rock salt, whereby the gold particles diffuse to atomic surface steps which become visible in this way. (See e.g. Picht and Heydenreich, 1966.)

References

Dislocation Theory

Bueren, H. G. van: Imperfections in crystals. Amsterdam: North Holland Publ. Co. 1961.

Cottrell, A. H.: Dislocations and plastic flow in crystals. Oxford: Clarendon Press 1953.

Fiedel, J.: Dislocations. Oxford: Pergamon Press 1964.

Hirth, J. P., Lothe, J.: Theory of dislocations. New York: McGraw-Hill 1968.

Nabarro, F. R. N.: Theory of crystal dislocations. Oxford: Clarendon Press 1967.

Read, W. T.: Dislocations in crystals. New York: McGraw-Hill 1953.

Seeger, A.: Moderne Probleme der Metallphysik. Berlin-Heidelberg-New York: Springer 1966.

— Handbuch der Physik, Bd. 7/1, S. 383. Berlin-Heidelberg-New York: Springer 1953.

Weertman, J., Weertman, J. R.: Elementary dislocation theory. New York: McMillan 1964.

Symposia on Dislocation Theory

Dislocation dynamics. Edit. by: Rosenfield, A. R., Hahn, G. T., Bement, A. L., u. Jaffee, R. I. New York: McGraw-Hill 1968.

Dislocations in solids. Discussions Faraday Soc. No. 38 (1964).

Fundamental aspects of dislocation theory. Edit. by: J. A. Simmons, R. deWit, R. Bullough. National Bureau of Standards. Special Publication 317. Washington, D.C. 1970.

Physique des dislocations. Paris: Presses Universitaires de France 1967.

Grain Boundaries and Phase Transformations

Amelinckx, S., Dekeyser, W.: Solid state physics, vol. 8, p. 325. London: Academic Press 1959.

Christian, J. W.: The theory of transformations in metals and alloys. Oxford: Pergamon Press 1965.

McLean, D.: Grain boundaries in metals. Oxford: Clarendon Press 1957.

Wayman, C. M.: Introduction to the crystallography of martensitic transformations. New York: McMillan 1964.

Special References

Section:

2.2-2.3 International tables for X-ray crystallography. Birmingham, England: Kynoch Press 1962.

4.1 Billington, D. S.: Radiation damage in solids. Proceedings of the Internat. School of Physics. "Enrico Fermi" XVIII Course. Italian Physical Society. New York: Academic Press 1962.

— Crawford, J. H., Jr.: Radiation damage in solids. New Jersey: Princeton University Press 1961.

Damask, A. C., Dienes, G. J.: Point defects in metals. New York: Gordon and Breach 1963.

Dienes, G. J., Vineyard, G. H.: Radiation effects in solids. New York: Interscience Publishers Inc. 1957.

Lattice defects in quenched metals. Edit. by: Cotterill, R. M. J., Doyama, M., Jackson, J. J., Meshii, M.: London-New York: Academic Press 1965.

Strumane, R., Nihoul, J., Gevers, R., Amelinckx, S.: The interaction of radiation with solids. Proceedings of the Internat. Summer School on Solid State Physics held at Mol/Belgium. Amsterdam: North Holland Publishing Co. 1964.

Vacancies and interstitials in metals. Edit. by: Seeger, A., Schumacher, D., Schilling, W., Diehl, J., Amsterdam: North Holland 1970.

5.2 Bilby, B. A., Bullough, R., Smith, E.: Proc. Roy. Soc. (London), Ser. A **231**, 263 (1955).

6.3 Weertman, J.: Phil. Mag. **11**, 1217 (1965).

6.5 Brown, L. M.: Phil. Mag. **10**, 441 (1964).

7.5 Hale, F. K., Brown, H. M.: Fourth European Regional Conference on Electron Microscopy, Rom 1968, vol. 1, p. 409.

8.3 Bollmann, W., Perry, A. J.: Phys. stat. sol. **24**, K141 (1967).

Thompson, N.: Proc. Phys. Soc. (London) B **66**, 481 (1953).

8.4 Bollmann, W.: Acta Met. **9**, 972 (1961).

Christian, J. W., Swann, P. R.: Stacking faults in metals and alloys (AIME-Conf. on "Alloying Effects on Concentrated Solid Solutions"). Cleveland, Ohio 1963.

Gedwill, M. A., Altstetter, C. J., Wayman, C. M.: Trans. AIME **230**, 453 (1964).

Hirth, J. P. (Symposia, ref. 3).

Seeger, A.: Z. Metallkunde **44**, 247 (1953); **47**, 653 (1961).

9.1 Bollmann, W.: Phil. Mag. **7**, 1513 (1962); — Discussions Faraday Soc. No. 38, 26 (1964).

9.3 Hale, F. K., Brown, H. M. (see ref. 7.5).

9.5 Thompson, N. (see ref. 8.3).

10.1 Damask, A. C., Dienes, G. J.: (see ref. 4.1).

Hirsch, P. B., Silcox, J. Smallman, R. E., Westmacott, K. H.: Phil. Mag. **3**, 897 (1958).

Hornstra, J.: Acta Met. **10**, 987 (1962).

Yoshida, S.: Kiritani, M., Shimomura, Y.: J. Electronmicroscopy **12**, 146 (1963).

11.8 Bailey, J. E.: Phil. Mag. **5**, 833 (1960).

Bollmann, W.: J. Inst. Metals **87**, 439 (1959).

Merklen, P., Furubayashi, E.: C. R. Acad. Sci. Paris **268**, 2159 (1969).

12.2 Ranganathan, S.: Acta Cryst. **21**, 197 (1966).

12.3 Bollmann, W.: Phil. Mag. **16**, 363, 383 (1967).

13.1 Fletcher, N. H.: Phil. Mag. **16**, 159 (1967).

Gleiter, H. (see ref. Section 14.2).

Ishida, Y., Hasegawa, T., Nagata, F.: J. Appl. Phys. **40**, 2182 (1969).

Levy, J.: Phys. stat. sol. **31**, 193 (1969).

13.2 Bollmann, W., Nissen, H.-U.: Acta Cryst. A **24**, 546 (1968).

13.3 Bollmann, W., Perry, A. J.: Phil. Mag. **20**, 33 (1969).

13.5 Schober, T., Balluffi, R. W.: Phil. Mag. **20**, 511 (1969).
— Phil. Mag. **21**, 109 (1970).

14.2 Bollmann, W.: (Symposia, ref. 3).
Gleiter, H.: Acta Met. **17**, 565 (1969a); **17**, 858 (1969b).
Marcinkowski, M. J.: (Symposia, ref. 3).
Sleeswyk, A. W.: Physique des dislocations (Ref. Symposia), p. 78.

14.3 Bollmann, W.: Dislocation dynamics (Ref. Symposia), p. 275.

A1 Z.B. International tables (Ref. 2.2).

A2 Oster, G.: The science of moiré patterns. Barrington, New Jersey 08007, U.S.A.: Edmund Scientific Co. (with two sets of moiré-models).

A3 Amelinckx, S.: The direct observation of dislocations. Solid state physics 6. New York: Academic Press 1964.
Brandon, D. G.: Modern techniques in metallography. London: Butterworths 1966.
Heidenreich, R. D.: Fundamentals of transmission electron-microscopy. New York: Interscience Publishers 1964.
Hirsch, P. B., Howie, A., Nicholson, R. B., Pashley, D. W., Whelan, M. J.: Electronmicroscopy of thin crystals. London: Butterworths 1965.
Müller, E. W., Tsong, T. T.: Field ion microscopy. New York-London-Amsterdam: Elsevier 1969.
Picht, J., Heydenreich, J.: Einführung in die Elektronenmikroskopie. Berlin: VEB Verlag Technik 1966.
Saada, G.: Microscopie electronique des Lames minces cristallines. Paris: Masson 1966.

Sources of Illustrations

Thanks are due for permission of reproduction of figures to the following publishers and Journals:

Books

New York: McGraw-Hill: Dislocation dynamics. Edit. by A. R. Rosenfield *et al.*, 1968, Figs. 13.4/1; 14.3/1, 2, 3, 4.

Utrecht: A Oosthoek's Uitgeversmaatschappij NV: Symmetry aspects of M. C. Escher's periodic drawings by Caroline H. MacGillavry, 1965 (© International Union of Crystallography). Fig. 2.2/1.

New York: Reinhold Publishing Corporation: The encyclopedia of microscopy. Edit. by A. L. Clark, 1961, Figs. 3.2/2, 3, 4; 3.9/1, 2; 3.10/1; 5.3/1; 5.4/1, 2.

Berlin-Heidelberg-New York: Springer: Einführung in die Mineralogie by C. W. Correns, 1968, Fig. 132/2.

Washington D.C.: National Bureau of Standards, special Publication 317. Fundamental Aspects of Dislocation Theory. Edit by J. A. Simmons, R. deWit, R. Bullough. 1970. Figs. 12.3/4; 12.8/2.

Journals

Acta Crystallographica: 13.2/1, 3, 5, 6, 7, 8.
Acta Metallurgica: Figs. 8.4/3, 4, 5, 6, 7, 8.
The Faraday Society: Figs. 11.1/1; 11.4/1; 11.6/1; 11.9/1; 11.10/1, 2.
Journal of the Institute of Metals: Figs. 11/8/1, 2, 3, 4, 5, 6.
The Philosophical Magazine: Figs. 9.1/1; 9.3/1, 2, 3, 4; 10.1/1; 11.4/2; 11.5/1; 12.5/1; 12.6/1, 2; 12.8/1, 4, 5; 12.10/1; 12.11/1; 13.3/2; 13.5/1, 3.
Zeitschrift für Kristallographie: Fig. 3.11/1.

Subject Index

The page numbers in *italics* indicate principal references

Made in the USA
Monee, IL
13 September 2026

62973770R00149